THE CLIMATE CHANGE EDUCATIONAL PARTNERSHIP: CLIMATE CHANGE, ENGINEERED SYSTEMS, AND SOCIETY

A REPORT OF THREE WORKSHOPS

Rachelle D. Hollander, Editor

Frazier F. Benya, Co-Editor

Cameron H. Fletcher, Co-Editor

NATIONAL ACADEMY OF ENGINEERING
OF THE NATIONAL ACADEMIES

THE NATIONAL ACADEMIES PRESS
Washington, D.C.
www.nap.edu

THE NATIONAL ACADEMIES PRESS 500 Fifth Street, NW, Washington, DC 20001

In September 2010, the Center for Engineering, Ethics, and Society at the National Academy of Engineering began working with four other partners on a Climate Change Educational Partnership (CCEP) Phase I planning grant from the National Science Foundation about "Climate Change, Engineered Systems, and Society." The partners were Arizona State University (ASU), the Boston Museum of Science (MOS), the Colorado School of Mines (CSM) and the University of Virginia (U Va) Much of the coordination and communication over the course of the grant occurred in scheduled monthly (and sometimes more frequent) telephone conferences. Five in-person meetings over the course of the grant were critical to identifying and addressing issues, developing a sense of common purpose, and engaging new partners; three of these meetings included public workshops. This report summarizes results from the workshops. Appendices include related materials.

(1) October 10-11, 2010: Initial meeting of partners at ASU, to review proposed activities and coordinate plans.

(2) June 6-8, 2011: Project planning meeting and open workshop on Climate, Technology, and Society at the National Academies' Beckman Center.

(3) October 17-19, 2011: Project planning meeting and open workshop on Networking Educational Priorities for Climate, Engineered Systems, and Society, at the House of Sweden in Washington, DC.

(4) January 10-11, 2012: Meeting of expanded partnership to plan for implementation project for education in climate change, engineered systems, and society, at CSM.

(5) January 28-30, 2013: Final project meeting – planning and open workshop at ASU.

This publication has been reviewed according to procedures approved by the National Academy of Engineering report review process. Publication of attributed work signifies that it is judged a competent and useful contribution worthy of public consideration, but it does not imply endorsement of conclusions or recommendations by the National Academy of Engineering. The interpretations and conclusions in such publications are those of the authors and do not purport to represent the views of the council, officers, or staff of the National Academy of Engineering.

This project was supported by Contract/Grant No. 1043289 between the National Academy of Sciences and the National Science Foundation. Any opinions, findings, conclusions, or recommendations expressed in this publication are those of the author(s) and do not necessarily reflect the views of the organizations or agencies that provided support for the project.

International Standard Book Number 13: 978-0-309-31275-2
International Standard Book Number 10: 0-309-31275-2

Additional copies of this report are available from the National Academies Press, 500 Fifth Street, NW, Keck 360, Washington, DC 20001; (800) 624-6242 or (202) 334-3313; *http://www.nap.edu.*

THE NATIONAL ACADEMIES

Advisers to the Nation on Science, Engineering, and Medicine

The **National Academy of Sciences** is a private, nonprofit, self-perpetuating society of distinguished scholars engaged in scientific and engineering research, dedicated to the furtherance of science and technology and to their use for the general welfare. Upon the authority of the charter granted to it by the Congress in 1863, the Academy has a mandate that requires it to advise the federal government on scientific and technical matters. Dr. Ralph J. Cicerone is president of the National Academy of Sciences.

The **National Academy of Engineering** was established in 1964, under the charter of the National Academy of Sciences, as a parallel organization of outstanding engineers. It is autonomous in its administration and in the selection of its members, sharing with the National Academy of Sciences the responsibility for advising the federal government. The National Academy of Engineering also sponsors engineering programs aimed at meeting national needs, encourages education and research, and recognizes the superior achievements of engineers. Dr. C. Daniel Mote Jr. is president of the National Academy of Engineering.

The **Institute of Medicine** was established in 1970 by the National Academy of Sciences to secure the services of eminent members of appropriate professions in the examination of policy matters pertaining to the health of the public. The Institute acts under the responsibility given to the National Academy of Sciences by its congressional charter to be an adviser to the federal government and, upon its own initiative, to identify issues of medical care, research, and education. Dr. Victor J. Dzau is president of the Institute of Medicine.

The **National Research Council** was organized by the National Academy of Sciences in 1916 to associate the broad community of science and technology with the Academy's purposes of furthering knowledge and advising the federal government. Functioning in accordance with general policies determined by the Academy, the Council has become the principal operating agency of both the National Academy of Sciences and the National Academy of Engineering in providing services to the government, the public, and the scientific and engineering communities. The Council is administered jointly by both Academies and the Institute of Medicine. Dr. Ralph J. Cicerone and Dr. C. Daniel Mote Jr. are chair and vice chair, respectively, of the National Research Council.

www.national-academies.org

Center for Engineering, Ethics, and Society Advisory Group

JOHN AHEARNE, Sigma Xi, The Scientific Research Society, Chair
ALICE AGOGINO, University of California, Berkeley
STEPHANIE J. BIRD, Ethics Consultant and Co-Editor of Science and Engineering Ethics
GLEN DAIGGER, CH2M HILL
GERALD E. GALLOWAY, JR., University of Maryland, College Park
DEBORAH JOHNSON, University of Virginia
WILLIAM KELLY, American Society for Engineering Education
FELICE LEVINE, American Educational Research Association
MICHAEL LOUI, University of Illinois at Urbana-Champaign
CHRIS SCHAIRBAUM, Texas Instruments, Inc.
CAROLINE WHITBECK, Case Western Reserve University
WILLIAM WULF, University of Virginia

Center Staff

RACHELLE D. HOLLANDER, Center Director
FRAZIER BENYA, Program Officer
SIMIL RAGHAVAN, Associate Program Officer
VIVIENNE CHIN, Administrative Assistant

Climate Change Educational Partnership (CCEP) Planning Grant: Climate Change, Engineered Systems, and Society External Advisory Board

DAVID DANIEL, President, University of Texas – Dallas
HELENE HILGER, Professor and Director, Civil and Environmental Engineering, IDEAS Center, University of North Carolina – Charlotte
JIM MCCARTHY, Professor, Biological Oceanography, Harvard University
RICKY ROOD, Professor, Atmospheric, Oceanic, and Space Sciences, University of Michigan – Ann Arbor
BOB ROTH, CEO, Big Green Zero
WALTER STAVELOZ, Director, International Relations, Association of Science-Technology Centers (ASTC)

Acknowledgments

This publication has been reviewed, in draft form, by individuals chosen for their diverse perspectives and technical expertise, in accordance with procedures approved by the National Academies. The purpose of this independent review process is to provide candid and critical comments to assist the committee and NAE in making its published reports as sound as possible and to ensure that the report meets institutional standards for objectivity, evidence, and responsiveness to the study charge. The reviewers' comments and the draft manuscript remain confidential to protect the integrity of the deliberative process. We wish to thank the following individuals for their reviews of this report:

Sarah Bell, Senior Lecturer, University College London
Glen Daigger, Senior Vice President and Chief Technology Officer, CH2M Hill
William Kelly, Director of External Affairs, American Society for Engineering Education
Jen Schneider, Associate Professor, Boise State University
David Sittenfeld, Program Manager, Boston Museum of Science

Although the reviewers listed above provided many constructive comments and suggestions, they were neither asked to endorse the views expressed in the report nor did they see the final draft of the report before its public release. The review was overseen by Gerald E. Galloway, Glenn L. Martin Institute Professor of Engineering, University of Maryland, College Park, who was appointed by NAE to ensure that an independent examination of this report was carried out in accordance with institutional procedures and that all review comments were carefully considered. Responsibility for the final content of this report rests entirely with the editors and the NAE.

In addition to the reviewers, CEES wishes to thank the project staff. Vivienne Chin managed the project's logistical and administrative needs, making sure the workshops ran efficiently and smoothly. NAE senior editor Cameron H. Fletcher edited the chapters for this publication, written by CEES director Rachelle Hollander and program officer Frazier F. Benya. Hollander and Benya also worked with the partnership partners to plan and carry out the workshops, and they coordinated and led the efforts to prepare project reports to NSF. Hollander led the editorial process for this publication and the response to review. CEES associate program officer Simil Raghavan assisted in preparing the manuscript for review. She interviewed project participants and developed several videos about the project for the Online Ethics Center website. NAE senior program officer Janet Hunziker managed the review process. CEES director Rachelle Hollander was the Principal Investigator for the grant; she oversaw the project from start to finish.

Finally, the four co-Principal Investigators for the award were instrumental in the creation of the partnership, the development of all project activities and resources, and the preparation and submission of reports to the National Science Foundation, which provided project funding. They are Clark Miller, associate professor, School of Politics and Global Studies, Arizona State University; Paul Fontaine, vice president of education, Museum of Science, Boston; Juan Lucena, professor, Liberal Arts and International Studies, Colorado School of Mines; and Deborah Johnson, professor of applied ethics, University of Virginia. These co-Principal Investigators were greatly assisted by project team members Jason Delborne, Jon Leydens, Junko Munakata Marr, Jen Schneider, and Kathryn Johnson from Colorado School of Mines; Joseph Herkert, Chad Monfreda, and Sharlissa Moore from Arizona State University; David Rabkin and David Sittenfeld from the Boston Museum of Science; Laura Sasso, David Slutsky, Andreas Clarens, and Michael Rodemeyer from University of Virginia, and Liz Cox from Red Rocks Community College.

Contents

Chapter 1
INTRODUCTION AND OVERVIEW

In September 2010, the Center for Engineering, Ethics, and Society at the National Academy of Engineering began working with four other partners on a Climate Change Educational Partnership (CCEP) Phase I planning grant from the National Science Foundation about "Climate Change, Engineered Systems, and Society." The partners were Arizona State University (ASU), the Boston Museum of Science (MOS), the Colorado School of Mines (CSM) and the University of Virginia (U VA).

The CCEP project focused on defining and characterizing the societal and pedagogical challenges posed by the interactions of climate change, engineered systems and society, and identifying the educational efforts that a network could use to enable engineers, teachers, students, policymakers, and the public to meet the challenges. For instance, improving public infrastructure requires attending to transit, waste, energy, water, and buildings as integrated systems. It also requires addressing the need for changes in engineering education to consider complex systems rather than individual technical problems, and the need for improved communications between public as well as private sector agencies.

Societies develop engineered systems to address or mediate climate-related problems, such as drought, sea-level rise or wildfire control; the mediation involves public trust, public engagement, and governance. In these efforts, societies also decide—intentionally or implicitly—questions of justice and sustainability, such as what areas will receive mediation measures, what types of measures will be used, and what levels and kinds of local impacts are tolerated. The project also aimed to build awareness of these complexities among a diverse set of communities affected by climate change and engineered systems and to engage the communities in addressing these challenges.

Over the course of the grant, the CCEP planning project on Climate Change, Engineered Systems, and Society held three workshops on the interactions of climate change with engineered systems in society and the educational efforts needed to address them.[1] The first workshop provided the partners with an introduction to the varied social and technical dimensions found in the relationships among climate, engineered systems, and society. These systems include social as well as technological factors and they both influence climate and are affected by it in positive and negative ways. For instance, water systems adequate under many climate conditions will be inadequate in others and inadequacy will depend to some degree on societal capabilities, perceptions, and expectations. The legitimacy of these expectations involves issues of governance, public trust, and equity of access to an essential resource.

The second workshop built on the common language developed in the first. It allowed the partners to expand involvement in the project to include representatives from community and tribal colleges, professional societies and business. It examined the opportunities and challenges for formal and informal education, particularly in engineering classrooms and science museums, to prepare students and citizens to address these issues. Presentations and discussions described the technical, societal, and ethical considerations that must be present in engineering education and societal decision making, if engineers and citizens are to be able to make informed choices.

The third workshop allowed the partners to broaden further the discussion and the audience. It solicited participation from government officials, Native American tribal representatives, professional society

---

[1] NSF awarded 15 CCEP phase I planning grants. The phase II competition made five awards for implementation projects. The partnership did not receive NSF funding for the second phase of this project. The Phase I CCEP award ends on August 31, 2014.

leaders, as well as educators, artists, scientists, and engineers who are developing programs that can manage change and educate students and citizens in ways that foster their leadership skills. The workshop focused the discussion primarily on infrastructure systems but also included cases that involved energy, agricultural/ecological, and manufacturing systems. Additionally, this final workshop modeled the range of interests and viewpoints that need to be represented in societal decision making processes about climate change, engineered systems, and society.

All three workshops attendees included leading academic researchers, from climate and earth sciences, engineering, ethics, science and technology studies, environmental science, and science and engineering education, and leaders from science museums specializing in informal science education. At the second workshop three additional communities were engaged: faculty from Native American and tribal colleges, representatives from professional engineering societies in the US and Canada, and practicing engineers and managers from corporate engineering firms, including Lockeed Martin, DuPont, General Electric, and CH2M Hill. At the final workshop, designed to engage a wide range of communities, all previous communities participated along with three new ones: (1) well known artists working to communicate climate science and environmental impacts through art; (2) leaders on Native American policy including the director of the US Department of Energy's Office of Indian Energy Policy and Programs, the director of the American Indian Policy Institute at Arizona State University, and the director of the Institute for Tribal Environmental Professionals; and (3) local and federal government officials ranging from the assistant director for climate adaptation and assessment at the White House Office of Science and Technology Policy to the county manager for Washoe County, Nevada.

This report summarizes the workshop presentations and discussions as they explored the project themes, from a variety of perspectives. This information may be useful to engineers, educators, corporate leaders, local and regional officials, members of professional societies, and others in their efforts to understand and address the challenges of climate change and its societal impacts.

NAE President's Perspective

Charles M. Vest, then president of the NAE, addressed the project team members and attendees at the second workshop. He spoke about his disappointment in the lack of US political and corporate leadership on climate change. Commenting on the earthquake and tsunami in Japan, which occurred just months before the workshop, he warned that this is an example of the kind of infrastructure disaster and vulnerability the world could experience as a result of climate change, and that the events in Japan should be a learning experience. Part of our opportunity and obligation regarding climate change, he said, is to minimize the probability that these natural disasters will have such devastating impacts. The international political response to the Fukushima nuclear plant incident showed that political will needs to be built now, rather than in times of crisis, to minimize the risks and impacts of disasters.

The Fukushima results also demonstrate opportunities and challenges in science and engineering education. For the CCEP project, Dr. Vest mentioned an NAE study on the use of engineering as an integrating factor for how to teach math and science at the elementary and secondary education levels.[2] It is hoped that the subject matter and discussions of the CCEP project will inform other NAE programs on issues in engineering education, especially as they relate to climate change, engineered systems, and society.

---

[2] Toward Integrated STEM Education: Developing A Research Agenda. The project report, *STEM Integration in K–12 Education*, was released in March 2014.

Workshops 1 and 2: Characterization of Issues

The first workshop, on "Climate, Technology, and Society," held in June 2011, included discussion of the social and technical facets of engineered responses to climate and their implications for governance, social justice, sustainability, and public trust and engagement. Speakers presented results from research on climate and its interaction with engineered systems, understood as *sociotechnical systems*. They examined adaptation, mitigation, and geoengineering as components in climate and engineered systems. (Session summaries are in chapters 2 and 3.)

Although the research literature addresses sustainability questions, sustainability is generally considered an environmental rather than a social issue, and relatively few articles consider or critically explore issues of social sustainability. The presenters explored the ways in which scientific, engineering, political, and social interventions and priorities can, do, and should influence the interactions of climate, engineered systems, and society, and how these influences are likely to affect the success of programs and proposed changes.

Discussions about responding to climate change often involve projections about the potential costs and benefits of various strategies. Going beyond cost-benefit analysis, speakers probed the social justice dimensions of these options—e.g., the kinds and distribution of potential benefits, costs, risks, and harms associated with them—and related considerations of governance, sustainability, and public engagement and trust. (See chapter 4.) There was general agreement that information at the national or international level tended to be abstract and speculative, whereas discussion about local initiatives tended to be pragmatic, inclusive, and involve grassroots participation.

If engineers and the public are to be prepared to address these issues, new and expanded educational programs must be provided. In considering the implications for education at the second workshop, "Networking Educational Priorities for Climate, Engineered Systems, and Society," held in October 2011, there was an effort to identify effective educational interventions, spanning undergraduate engineering curricula, community and tribal college programs, K–12 education, informal education and public engagement, public policy education, and outreach, dissemination, and special projects. Chapters 5 and 6 of the report summarize the results of the second workshop.

Effective undergraduate interventions may include innovations to improve (1) the integration of climate change issues and engineered system approaches to them in engineering curricula and (2) scale-up across multiple institutions. The overarching premise is that engineers should be specifically trained to address climate change issues. From a corporate perspective, for example, engineers should have an understanding of the association between their practice and its implications for climate, engineered systems, and society. Businesses that employ engineers can provide useful information about the demands and technological and organizational challenges they see ahead, and about the principles, skills, and experiences needed for future engineers to meet the challenges of climate change.

Effective informal educational efforts must focus on relevant topics and use engaging formats and materials, which can be adapted to different contexts. Professional societies also are developing initiatives relevant to educational priorities for climate, society, and technology. And there is a role for science and technology centers, which can communicate multifaceted information and present a model for engaging both the general public and school-aged audiences.

Workshops 1 and 2: Opportunities and Needs in Formal and Informal Education

A number of presentations revealed opportunities for the implementation or development of educational interventions and innovations; for example, informal education activities can address the needs of teachers and students in secondary schools, web-based initiatives can be a forum for online collaborations and shared resources, and case studies can be used to illuminate issues of climate, engineered systems, and society.

Informal evidence from science centers suggests that the topic of climate change, engineered systems, and society has strong appeal for engaging a lay audience in examining the important relationship between the natural and human-made worlds.

A variety of approaches can be used to encourage formal and informal education on climate change, engineered systems, and society, such as case studies, courses, degrees, modules, exhibits, extracurricular activities, interdisciplinary collaboration, prizes, institutes, forums, and specialized training involving rethinking concepts, theories, and worldviews.

Research is growing about climate change and engineered systems, but two important gaps persist. First, climate change remains largely absent in engineering curricula, with the exception of offerings in renewable energy engineering. Second, few if any materials fully engage the integration of climate, engineering, and human systems. To promote effective learning, new materials, particularly new case studies, are needed.

Case studies are likely to have the most general applicability; they can be tailored to diverse learning contexts and audiences by adapting them with local data and information as well as problems of professional relevance, and by crafting educational packages for four target audiences: adults and youth, communities and community leaders, undergraduates, and engineering and technical students. The project partners identified four areas for preliminary cases that would allow for region-specific adaptations—ports and sea level rise, urban heat islands, urban water and wastewater systems, and engineered river systems—as well as a globally significant example: the Panama Canal. Cases in these areas take advantage of ongoing activities and interests of the network partners.

Pedagogical needs arise at multiple levels. Faculty interviews[3] about constraints and possibilities for addressing the selected issues in the engineering curriculum indicate that they perceive both threats and opportunities—institutional (budget and faculty availability, possibilities for collaborations and funding), curricular (changing course content), epistemic (the indeterminate and unpredictable relationship of climate change to design criteria for engineered systems), and political (industry affiliations and the political sensitivities of climate change).

In informal education, the general public and community leaders need knowledge about local vulnerabilities and venues in which to discuss their implications. Community leaders need to be able to engage stakeholders in productive problem solving. In formal education, students need to recognize how natural, social, and technological forces together influence the future and to learn to work in interdisciplinary contexts. Areas of technical specialization require knowledge in subjects such as design and resilience. Training for all professions should include information about professional responsibilities. Effective dissemination and adoption of new methods and resources will require educationally appropriate materials and professional development opportunities for faculty members and specialists in informal learning as well.

[3] The CSM partners conducted these interviews as part of the phase I project, to determine faculty responses to suggestions for curricular innovation.

Efforts to meet these ends should incorporate (1) a focus on adults and college students, (2) the identification and design of integrated learning resources around key vulnerabilities of US infrastructure and energy and manufacturing systems to climate change, (3) the design of resources based on the latest insights into STEM learning from the learning sciences, and (4) the establishment of professional and online educational networks to expand educational opportunities. In addition, policymakers and others should be engaged to promote attention to and understanding of the importance of these approaches for national educational initiatives.

Themes Articulated at the Capstone Workshop

At the capstone workshop in January 2013, on "Climate Change and America's Infrastructure: Engineering, Social, and Policy Challenges," experts and stakeholders presented perspectives on vulnerabilities in engineered systems, the role of art in communicating with the public, uncertainty, local impacts, and Native American experiences, among others. Summaries of the presentations and discussions are integrated in all of the report chapters, with chapter 7 containing the contributions from representatives of professional societies, business, and Native American tribes to the capstone.

In the session discussions, participants indicated that in seeking to address engineering, social, and policy challenges associated with infrastructure, it is important to take into account the implications of regional climate variability, strengths and weaknesses in interconnected infrastructures, and the need for integrated action to deal with increasing potentials for risks and disasters to engineered systems in the face of climate change. These are often linked with issues of policy, governance, justice, and human rights, for which government action and responsiveness are required, although educational and community programs can also be engaged to identify problems and propose solutions.

Workshop speakers and participants demonstrated a deep appreciation for the relationships between climate and society and for the difficult challenges to engineered systems that experts and communities need to face together. Infrastructure, including energy and manufacturing systems, needs to be planned and built to last many years. Furthermore, as it often takes many years to build and complete, during which social and environmental conditions change, planning needs to take that into account to ensure that the infrastructure is resilient. This understanding is important to both experts—urban planners, regulators, elected officials, and engineers, among others—and the public.

Two examples at the capstone examined infrastructure vulnerability and ways to engage with policymakers and the public: Florida's vulnerability to sea level rise, and approaches to climate-related decisions when the science is uncertain. Panels on vulnerability featured local government officials, both legislative and executive, and artists who had initiated projects to illustrate climate change impacts, with slides and videos of their efforts. The panel on uncertainty brought together scientists, engineers, and officials from the private and public sectors to discuss issues associated with Colorado River water resources. The panelists presented reports from studies and decision exercises that seek to address problems associated with increasing claims on resources and growing uncertainty about their availability.

Native American communities face particular issues associated with climate change ranging from substandard infrastructures to difficulties in maintaining their jurisdictional prerogatives. They also have some advantages insofar as they have a young population, opportunities to build from scratch and develop resources, and a deeply held concern about place and sustainability.

Several areas of particular interest emerged from the presentations and discussions throughout the workshop: the importance of focusing on decision-making processes, technical analysis, and educational priorities.

Decision-Making Processes

Many speakers and audience members agreed that all constituencies and stakeholders need to be involved in decision making about climate and infrastructure,[4] and that technical approaches need to be inclusive. Engineering efforts should engage professionals, operators, and managers with pertinent local knowledge and they should emphasize robustness, adaptive management, and the ability to respond positively to crisis and change.

Experts and local government officials in attendance observed that building social capital requires involving all stakeholders, including media and business as well as multiple agencies, perhaps multiple jurisdictions. Knowledge is distributed in all these groups; governance should increase social sustainability and reduce long-term social risks through inclusive efforts. Many at the conference agreed that advance preparation could establish social capital and thus take advantage of moments and targets of opportunity.

There was general agreement that more attention to questions of infrastructure, justice, and human rights is needed, especially because of human vulnerabilities and the length of time it takes to develop infrastructure. Objectives change over time and technological fixes that at first seem wonderful or even adequate may end up being of limited value. For example, as the levees were built for New Orleans, their height was not changed although the ground was demonstrably sinking during the project.

Technical Analysis

Participants noted the needs for flexibility in assumptions about climate and infrastructure change, sustained climate assessment over time, and development of process (not just physical) indicators (an example of a *process indicator* would be whether a city has considered its vulnerabilities).

Many agreed on the need to do "backward" analysis, rather than traditional risk assessment, to focus first on plans/scenarios and analysis of implications of their vulnerabilities, and then develop robust strategies that are good over a wide range of potential outcomes. Iterative analyses are necessary because of the immense uncertainties.

Educational Priorities

Engineering faculty at the workshop agreed that societal and ethical questions can be built into engineering classes, but not easily. Critical and systems thinking, social inclusion, and environmental justice are not part of the standard engineering curriculum. Training the next generation of engineers to consider questions of climate, infrastructure, and society is a very important step, but engineering faculty members need incentives to develop new courses or modules in these areas.

Audience members pointed out that science and technology centers and museums are trusted community resources and have developed popular and informative programs on climate, infrastructure, and society. These programs frame the issues, convene participants, and catalyze action. Visitors of all ages come to participate in forums and to view and interact with exhibits. These centers share resources and program results with their sister organizations around the country, and many at the conference agreed that they have a useful role to play in developing and continuing the nation's conversation on climate and America's infrastructure needs.

---

[4] The term "infrastructure" here includes energy and manufacturing systems as well as other civil works infrastructures.

Organization of the Report

This chapter provides an overview of the project, characterizing the challenges and approaches to address them identified in the series of public workshops. The chapters that follow are organized thematically, drawing from relevant presentations across workshops. Chapters 2, 3, and 4 draw on material and discussions from the first and final (capstone) workshops to identify and analyze the problems associated with the interactions of climate change, engineered systems, and society. Chapters 5 and 6 draw from the second and final workshops to identify and explore opportunities for formal and informal education in academic institutions and other community venues such as science museums. Chapter 7 summarizes perspectives from professional society, business and industry, local government, and Native American representatives, using material from the second and capstone workshops. The appendices contain the workshop agendas, the lists of project participants, and a summary of the results from the workshop evaluations.

This report presents a sociotechnical systems approach to engineering education and to broader societal consideration of responses to climate change. It identifies the technical, societal, and ethical issues that need to be addressed. As a factual summary of the contents of presentations and discussions at the workshops, it does not draw conclusions from the material. Rather, it presents the wide variety of perspectives and resources that need to be brought to bear on the topic. The final workshop in particular modeled the range of interests and viewpoints that need to be represented in societal decision making processes about climate change, engineered systems, and society. Analysis of the workshop evaluations, presented in Appendix C, demonstrates the success of these workshops in increasing understanding for the many different participants and provides suggestions for future activities.

Chapter 2
INTERACTIONS: DEFINING THE PROBLEMS

This chapter and the next two pull together material from the project's first and last (capstone) workshops that identified likely interactions among climate, engineered systems, and societies and the range of available responses. Project co-principal investigator Juan Lucena of the division of Liberal Arts and International Studies at the Colorado School of Mines (CSM), moderated the opening session, in which speakers described science perspectives, business and engineering perspectives, and public perspectives. Project team members Joseph Herkert at the Arizona State University School of Applied Arts and Sciences and Jason Delborne, CSM Liberal Arts and International Studies, provided overarching observations about the presentations, focusing on their implications for building a network and improving educational efforts and outcomes.[5]

Science Perspectives

James McCarthy, professor of biological oceanography at Harvard University, reviewed the evolution of climate science from 1980 to 2010, demonstrating that although the evidence for global warming has been increasing for many years (Figure 2-1), it has not translated into effective action. In the 1980s several workshops sponsored by the National Research Council (NRC) indicated growing concern; from those discussions came the creation of the Intergovernmental Panel on Climate Change (IPCC) in 1988. The UN Framework Convention on Climate Change was established in June 1992, at the UN Conference and Development in Rio de Janeiro, where President George H.W. Bush signed the treaty on behalf of the United States; in October 1992 the US Senate ratified it. By 1994 enough nations had ratified it for the treaty to come into force.

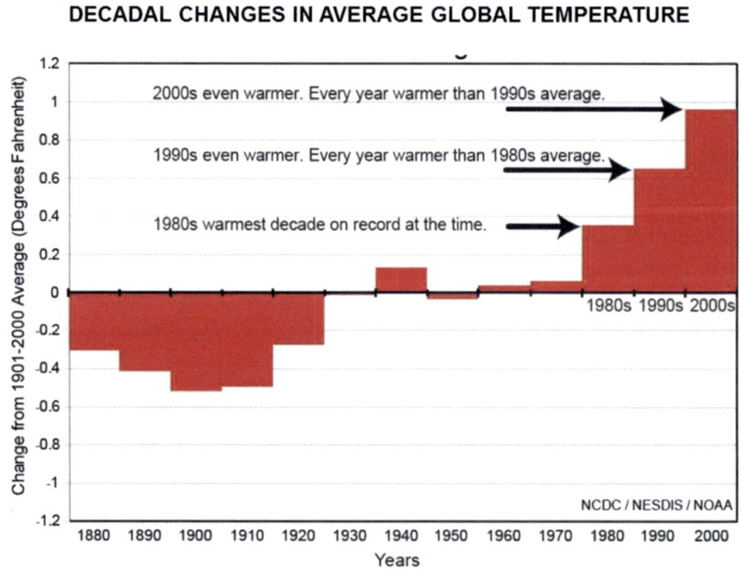

Figure 2-1 http://www.ncdc.noaa.gov/bams-state-of-the-climate/

The UN Framework Convention calls for the signatory nations to work together to reduce CO_2 emissions, but the emissions rate began to soar. Between 1995 and 2000, the evidence of unusual weather patterns and extreme events was consistent with theory on what to expect in a warmer world. But factors that promote increases in emissions—particularly from developing countries, which produce items consumed

[5] The agenda with links to slides from the speakers' presentations at the first workshop is available at http://www.nae.edu/Projects/CEES/57196/35146/60202/47874.aspx. The appendixes to this report provide the three workshop agendas and list the project investigators and staff.

by developed nations—continue to overcome concerns about the problems associated with those emissions because developed countries benefit greatly from this production.

McCarthy reported that much of what concerns people is the extremes: unusually cold periods become very rare, hot periods become hotter, and there are new record highs. When extreme highs become common, the coping range of humans (and other animals) is exceeded. McCarthy cited the heat waves of Paris in August 2003 and Russia in summer 2010, which caused some 15,000 and 55,000 deaths, respectively.

Complex parts of the climate are difficult to explain and are often seen as contradictions. People may point to the extreme cold in winter 2012 as a counterexample to global warming, but it is not in fact unrelated to the warming in the Arctic. New winter records for low ice cover in the Arctic create more open water and change the polar circulation so that cold air spills out of the Arctic and into some parts of southern Canada and the northern United States.

Social uncertainty—about what we humans will do, the choices we will make, and how they will affect whether we will have a low-, medium-, or high-emission future—further complicates the problem. President George W. Bush indicated in 2000 that he would sign the Kyoto protocol, but then changed his mind. None of the bills introduced in the US House and Senate has been enacted and the prospects are bleak. Pew public opinion polls show that the issue is a partisan one, and that people with a higher level of education are more likely to believe that human activity has caused global warming.

McCarthy concluded that there is absolutely no reason to believe that more definitive data and numbers or better explanations will lead people to agree on the severity of the problem or how to address it. But engineers can help point the way forward, by improving engineered systems that address climate as well.

Business and Engineering Perspectives

Jay Golden, director of the Center for Sustainability and Commerce at Duke University, posited that thinking about engineering and climate change requires consideration of manufactured goods and the expansion of the middle class: increases in population, urbanization, and the middle class go hand in hand with increased demands for services and engineered, manufactured products. He challenged a commonly accepted perspective that limits thinking about climate change to its impact on the built environment and said it is essential to recognize the influence of institutional drivers, private as well as public, on growing consumption.

Looking at the built environment *together with* the goods used in it (both products and services), transportation, and individual and organizational behaviors focuses attention on economic gains and savings from energy efficiency. This composite perspective also reveals the impact of climate change on the availability of resources, such as the bio-based products that will be needed to feed a much larger middle class living in cities; Golden calls this perspective *sociometabolic consumption*.[6] Engineers will have to work with industry, government, and nonprofit organizations to address these problems, he said.

---

[6] See http://www.uni-klu.ac.at/socec/inhalt/1928.htm. Some systems engineering programs focused on energy take as premises that socio-economic systems depend on a continuous throughput of materials and energy for their reproduction and maintenance like those of the metabolisms of biological organisms. An objective of these programs, such as the one identified in this footnote, can be to describe and analyse socio-metabolic patterns at different scales and identify points of intervention for guiding consumption in a more sustainable direction.

The next two generations will see the equivalent of about 1,000 megacities (i.e., with more than 10 million inhabitants), and the associated alterations occurring in the canopy (or the atmosphere in, above, and around them) from the physical engineered infrastructure and designs of urban dwellings in these cities will send a strong climate signal.

Dense urban environments also create issues of social justice, such as heat exhaustion and morbidity from heat-related illnesses, inequities in access to energy-efficient devices, and increases in crime and human health morbidity and mortality. How should these factors be incorporated in the education of future engineers?

Golden indicated that recent study of climate change finds that manufactured goods during their full life cycle contributed 50 percent of the problem. So focusing on built environments leaves out much of relevance in engineering fields such as mechanical, electrical, transportation, mining, and industrial. Results from these fields often add to sociometabolic consumption, and can exacerbate contributions from engineering to climate change. Changing this role requires understanding how the needs of clients influence the focus and outputs of engineering schools and knowing what is asked and expected of engineers when they leave school.

The product life cycle is a related engineering perspective that needs broadening, particularly to include attention to human behavior. Knowledge of the impacts of manufactured goods requires consideration of consumer use. If Apple computers are designed to be highly energy efficient but consumers leave them on all night, there may be no energy savings. Understanding the complexities of sociometabolic consumption calls for broader, more inclusive perspectives from engineers.

For a simple example of how consumer use affects climate change, Golden used the illustration of doing laundry. Simple design decisions have lowered the energy use associated with washing clothes: cold water washing and horizontal loading equipment can make significant inroads on the need for coal-fired power plants. Many people still believe that the biggest impact is from heating water, but mechanical drying has far greater impacts, particularly in the United States, where very little laundry is hung to dry.

Other important drivers are regulations and business-to-business initiatives; the latter are the largest global driver, through the Sustainability Consortium. Golden cited Walmart, which told its vendors that it will make purchases based on the full life cycle of the product, from manufacture to consumer use and postconsumer disposition. Companies such as Procter & Gamble and Unilever, with 30 percent of revenues tied to Walmart, and business associations with multinational members pay attention to such announcements. Thus retailers, manufacturers, and suppliers will have to be transparent as their products, including life cycle impacts, are audited.

Golden concluded by identifying other positive initiatives. One is open collaboration, such as Nike's work with others to make a new technology freely available; while another individual or company can't profit from that use, it can improve it and make a profit on the improvement. He noted the models at his university of a certificate program in engineered systems and sustainability, and sustainable energy fellowships linking theoretical with hands-on experience and examination of the political, economic, and social realities underlying sociotechnical change.

Public Perspectives

Ann Bostrom, dean of research and professor at the University of Washington School of Public Affairs, addressed three issues associated with public trust and engagement: (1) engineering risk assessments and engineering expertise; (2) lay risk perceptions and how they contrast with those of engineers; and (3) steps to better decisions, which involve engaging stakeholders.

Stereotypes about engineers being highly numerate and technically minded are borne out by some research showing that they do, in fact, prefer accuracy and lots of technical detail, she said. Lay people, on the other hand, have problems understanding probabilities and tend to make affective evaluations rather than numerical comparisons, using feeling, a rapid and automatic response, to assess the positive or negative quality of a stimulus. Everybody does this, but those who are numerate might check the details or do the calculations and then respond differently.

Decades of research on trust have led to several major models in the field. One takes a twofold approach to study how people decide whether to cooperate. First is social trust, where people look at morality or values information, including social and cultural similarities; if this information matches their values, the match engenders social trust in an interpersonal situation. Second, people evaluate information based on performance—their observations of what's happened, competence, and so on (Figure 2-2).

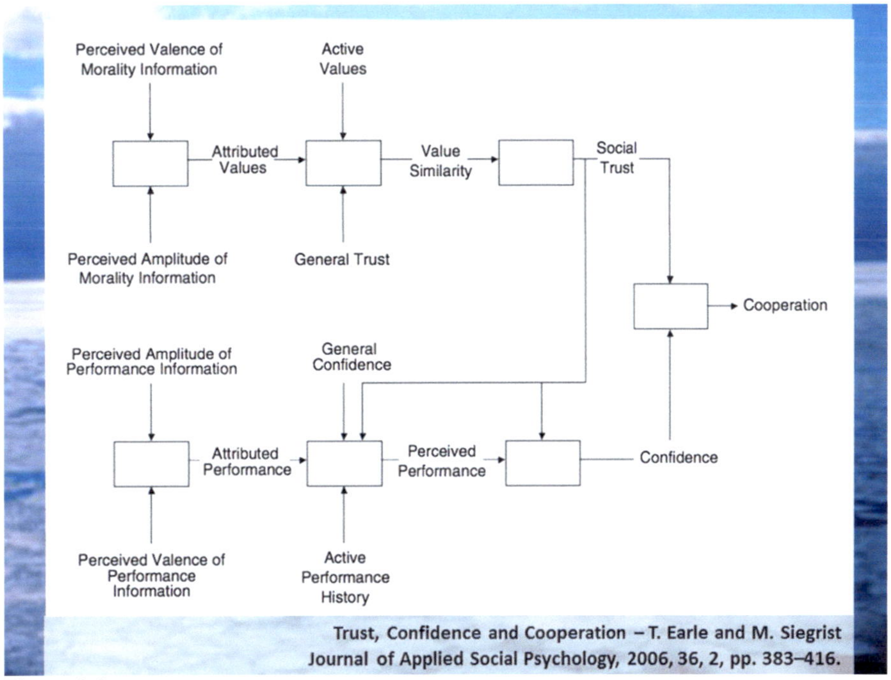

Figure 2-2

Whatever the technical information, however, people, even experts, will in most contexts turn to the affective information to decide on trustworthiness, and will be insensitive to the probabilities. So in efforts to address climate change, if people need to engage or cooperate, affect will trump numbers. Similarly, unlike technical risk assessments, lay risk assessments tend not to consider statistics and to compress probabilities, seeing lower probabilities as higher and higher as lower. They pay attention to different information, attending to and remembering problems based on their cognitive limitations and what's salient at the moment. Views depend on perceived threat and the perceived efficacy of actions to diminish it, attention, and other factors.

Furthermore, research on public opinions indicates a marked difference between measures along a scale of egalitarian to hierarchical and individualistic to communitarian. Asked about risk from global warming, people with individualistic and hierarchical values think it's substantially lower than do those with high egalitarian and communitarian scores. Some people contend, said Bostrom, that these differences underlie political differences in responses to climate change.

Experts must be aware of public processes and values if they want to assist in solving problems associated with climate and engineered systems. Bostrom and other researchers have developed mental models of how people think about hazardous processes and the systems such processes interact with, and how they understand causal processes. People tend to think of causal processes using metaphor and analogy, so they are heavily influenced by how the problem is framed and the analogies used. These results have been replicated nationally, with findings of both misconceptions and complex understanding about climate change mechanisms and solutions.

Despite these differences in understanding and perspective, there is considerable evidence that lay participation can produce better decision making. For example, members of local publics may have relevant expertise. There are several approaches to participative decision making but in all cases the first thing to remember is that affect trumps performance information. Further, it should be admitted that much important information about hazard-associated deaths, expenses, displacement, distribution of risk and benefits, is disputed; that there are methodological and ethical challenges to figuring out what people value and what should dominate in decision-making efforts; and that these are further complicated by uncertainty. Facilitated discussions of engineers with other parties can help address these issues.

Bostrom presented a "choice architecture" (or "nudges") for useful guidance in decision making:

- **In**centives for people to feel they are getting something for their choice
- **U**nderstanding how people see things
- **D**efaults: make sure the "do nothing" route is one of the best
- **G**ive feedback: investigate rejected options and experiment with them
- **E**xpect error: humans make mistakes, and well-designed systems allow for this
- **S**tructure complex choices: if it's difficult, break it down into easier chunks

In the question and answer period, Bostrom expanded the concept of moral information as social information about, basically, whether a person is "good" or "bad." Emphasizing shared values can help persuade people at least to consider what you're saying. She thought the use of cases was a good strategy for climate change education for engineers, but that findings on the effects of values on perceptions and choices of experts and nonexperts alike should inform the general understanding of engineers and nonengineers as well as engineering thought about designing, implementing, and assessing systems. Another important focus should be on the effectiveness of facilitated dialogue and decision engagement strategies.

Remarks on the Presentations

Delborne highlighted some session ideas that he believed provided useful information for the project. McCarthy's presentation demonstrated that increased knowledge and certainty needn't translate into action. How uncertainty is applied can have significant ramifications, as in the use of a smaller number for sea level rise. How engineers deal with uncertainty needs to be examined. Also relevant is the mismatch between scientific confidence in climate change and that in the media and public discourse, as is the question of which groups of citizens engineers and policymakers listen to.

Golden's talk was thought-provoking, Delborne said, about how the problem of climate change is defined. Focusing on cities as a built environment that experiences climate change in a more extreme way can motivate engineers; the challenge is to expand the network of those involved beyond civil and environmental engineers to perhaps mechanical engineers and others. What interventions would work to educate these other groups of engineers? Does bringing a broader variety of engineers into the network help or does it create different challenges?

Bostrom's remarks raised the question for Delborne of what cooperation the project aims to achieve. There's a big difference between getting public cooperation to follow expert advice, cooperation between experts and the public to make decisions together, and cooperation based on a more systematic and deliberative democratic process. It is also necessary to recognize the tension between lay and expert perspectives: people in general offer confused interpretations of technical information, whereas engineers have confidence that people know a lot and that participatory decision making is good, he said. Will people who cannot communicate well be discounted? There is value in the participatory process and in dialogue among experts, engineers, scientists, and the public, but those interactions can also give way to negative judgments from the public about engineers and from engineers about the public. Delborne concluded that it may be necessary to create a network that spans expert and lay communities, to pay attention to how they perceive each other and what kind of trust they build and how.

Herkert reiterated the point that engineers are not monolithic in their thinking or positions; like any group, they have a lot of commonalities *and* differences. He considered the three issues that are the focus of this project—ideas about enhancing the network, educational reforms, and the sociotechnical systems that comprise the interactions among climate, engineered systems, and society. First, he identified two takeaway messages for enhancing the network. One is to make sure it includes people with expertise who are also expert communicators; the other is to expand the notion of engineered systems, as suggested by Golden, to include manufacturing and manufactured products to include all engineers, regardless of academic discipline.

For educational reform, Herkert noted that Golden's talk indicates the need for more of a systems approach in engineering; some fields are embracing this approach quite readily while others are slower. Bostrom's presentation raised issues to be considered in communication between engineers and nonengineers. The results from research on risk perception, he said, may be hard for engineers to understand, but such understanding is critically important to the necessary dialogue between the two communities.

Last, the notion of sociotechnical systems is difficult to convey to engineers, and the project will have to address that difficulty. Herkert cited the Golden talk as a good example of how to convey the notion to engineers.

Further Perspectives on Interactions: The National Climate Assessment, Regional Impacts of Climate Change, and Infrastructure

Four talks at the capstone workshop focused on the findings of the National Climate Assessment, the implications of regional climate change given high climate variability, and strengths and weaknesses in interconnected infrastructures.[7]

Kathy Jacobs, assistant director for climate adaptation and assessment, White House Office of Science and Technology Policy, opened with a presentation titled "Engineering, Adaptation, and the National Climate Assessment." She discussed the challenges of climate change for engineered systems, the recent draft National Climate Assessment (ncadac.globalchange.gov), the role of ecosystem services and ecosystem-based approaches in engineering, the role of engineering in adaptation and resilience, and the importance of sustained assessment.

She made the following broad suggestions for (re)framing the role of engineering:

- Make sure you are solving the right problem.
- Engage a broad range of stakeholders and decision makers in collaborative, participatory processes to focus on solutions.
- Leverage existing systems, institutions, partnerships, and networks to build on existing capacity.

[7] Video and slides from the workshop are available at www.regonline.com/builder/site/tab2.aspx?EventID=1155563.

- Understand regional culture and its influence in decision making. Identify and engage trusted intermediaries who can assist with coordination.
- Advance coordination and evaluation processes based on shared learning and joint problem solving.

Moving from the national to the regional level, Daniel R. Cayan, researcher, Climate Atmospheric Science and Physical Oceanography (CASPO), Scripps Institution of Oceanography, spoke on "Regional Climate Change on Top of Already High Climate Variability." Using California as an example, he showed the implications of climate change for increases in days of extreme heat, numbers of forest fires, coastal flooding, and other untoward weather events. He concluded that

- Warming is already under way and projected to get worse.
- Along with warmer mean temperatures, extremes will intensify—heat waves will become hotter, longer, and occupy a broader season.
- Recent IPCC model projections for precipitation are scattered, but several simulations show moderate drying in the Southwest and increases in precipitation across the northern tier of the United States.
- Wildfire could become a greater threat.
- Climate warming projections, combined with recent global sea level rise (SLR) estimates, suggest increases along the West Coast sea levels of 0.5m to more than 1.5m by 2100.
- Tides, weather, and short period climate (such as increasing runoff from "warm storms" and decreased snowpack) will exacerbate SLR impacts.
- To plan and prepare for impacts, knowledge of regional and local details of climate, natural, and human systems matter greatly. Vulnerability assessments and downscaling are crucial.

The next two speakers focused on infrastructure. Thomas Wilbanks, corporate research fellow, Climate Change Science Institute, Oak Ridge National Laboratories, talked about "Climate Change and the Resilience of Interconnected Infrastructures," and Gerald Galloway, Glen L. Martin Institute Professor of Engineering, University of Maryland, gave the lunchtime address on "Climate Change, Engineering, Disasters, and Risk: It's Time to Do Something!"

Wilbanks pointed out that many consequences of climate change involve interactions among various kinds of built infrastructures and environments. Urban areas often hold special interest for cross-sectoral attention because their infrastructures are integrated (and because that is where most people live and vote, and where the financial and media centers are, for example). Critical cross-sectoral interactions are also issues at the regional (e.g., electricity infrastructures and communication infrastructures; transportation and waste disposal infrastructures) and national (national security) levels.

Risks of disruptive impact can be substantially reduced, he explained, by developing and implementing appropriate adaptation strategies based on information such as standards, codes, certification programs, and other practices that set rules for infrastructure; partnerships between the public and private sector; special attention to infrastructure that is near the end of its lifetime or performing poorly; and leadership and effective governance. Other elements of adaptation strategies often include the bundling of climate change responses with other development and sustainability issues, attention to financing, and efforts to spur innovation. Signs of progress include a number of bottom-up initiatives by US cities and attention to adaptation research needs for infrastructures.

Galloway identified the big picture in which scientists and engineers must operate today. The traditional approach assumed little change in climate and human behavior, operated within a narrow future, and stayed inside disciplinary stovepipes. Today's approaches require struggling with hundreds of possible climate and anthropogenic-driven scenarios, shared responsibilities, and adaptive, complementary efforts. He used a series of examples, as recent as Superstorm Sandy, to make this point.

Determination of risks and what to do about them requires many different people and entities, including scientists and engineers and their organizations, to assume responsibility. Risk is complex, it changes over time, people don't know they are at risk, and those responsible for communicating about it are not doing so effectively, he said. Structural and nonstructural interventions, including policy changes, need consideration. He contrasted the lack of direct US policy about risk with that of the United Kingdom and the Netherlands. The UK in 2005 issued guidance stating that the government will act proportionately and consistently in dealing with risks to the public, basing all decisions about risks on what best serves the public interest, with actions to be taken proportionate to the level of protection needed and targeted to the risk.[8] The Royal Netherlands Embassy in Washington DC issued a statement on September 16, 2008 indicating that the government had established an independent committee to issue advice as to how to improve the flood protection levels of all diked areas by a factor of 10 before 2050.

In Summary

Sessions in the first and capstone workshops of this Climate Change Educational Partnership Phase I project focused on examining the interactions between the phenomena from which the project took its name - "Climate Change, Engineered Systems, and Society." The presentations and discussions summarized here illuminated the nature and complexity of climate, engineered systems, and society as a sociotechnical system and discussed their implications for enhancing the network and educational reforms. The next chapter summarizes material from workshop sessions on interventions intended to address these complex phenomena.

---

[8] 2005. HM Treasury. Managing Risks to the Public: Appraisal Guidance.

Chapter 3
INTERVENTIONS: EXAMINING THE RANGE OF SOCIOTECHNICAL RESPONSES

The next session at the initial workshop concerned interventions to address the problems set forth in the earlier presentations (Chapter 2). Speakers and participants considered adaptation, mitigation, and geoengineering, placing them in their social contexts and recognizing that they pose challenges for social justice as well as governance, sustainability, and trust. They examined their policy and educational implications in light of the network's goals.[9] Junko Munakata Marr, associate professor of environmental science and engineering at the Colorado School of Mines (CSM), chaired the panel and introduced the topic and speakers.

Mitigation Strategies: Potentials and Problems

Ed Rubin, alumni professor of environmental engineering and science and professor of engineering and public policy and mechanical engineering at Carnegie Mellon University, opened with remarks on setting goals and determining effective options.

He mentioned four mitigation approaches: reduction in demand for energy-intensive goods and services, improvement in energy efficiencies, expanded use of low and zero carbon energy sources, and direct capture and sequestration of CO_2 from ambient air (a geoengineering approach). All of these methods are available to some extent in various forms; all have behavioral as well as technical components.

The focus of his presentation, however, was policy and education options. His discussion of policy options drew heavily on a 2010 NRC report, *Limiting the Magnitude of Future Climate Changes*, for which he served on the authoring panel. In addition to making policy recommendations, the panel considered issues of equity and environmental justice and the importance of flexibility in designing policies that are both durable and consistent.

Setting a Climate Change "Budget"

In recent years there has been some international political consensus on a roughly 2°C long-term increase in global temperatures. To set mitigation goals, the NRC panel considered what a safe amount of climate change would be and calculated how that would translate into atmospheric concentration, when change is likely to stabilize, and how that in turn establishes limits on the total amount of greenhouse gases that can be added to the atmosphere. The panel then sought to determine a reasonable allocation to the United States from a global "budget."

One key message of the study was the concept of a budget—people generally understand budgets. With 2050 as a timeframe, the United States would have about 40 years to meet the targets. Budgets from two recent studies[10] would establish targets of 50–80 percent reductions below recent levels. On the current

---

[9] The agenda with links to slides from the speakers' presentations at the first workshop is available at http://www.nae.edu/Projects/CEES/57196/35146/60202/47874.aspx. All the workshop agendas are available in Appendix A.

[10] Based on results from Energy Modeling Forum -22 (EMF, 2009) https://emf.stanford.edu/projects/emf-22-climate-change-control-scenarios; and National Academy of Sciences, National Academy of Engineering, and National Research Council. America's Energy Future: Technology and Transformation: Summary Edition. Washington, DC: The National Academies Press, 2009.

trajectory in the United States, that budget would very quickly be "spent" well before the middle of this century, meaning that the current pace of efforts to address climate impacts has to change, he concluded.

Technical, Policy, and Education Options

The human activities and two energy sources that give rise to most CO_2 are electricity and transportation (i.e., coal and oil), which are 75 percent of the domestic and international problem. So the climate problem requires tackling the two things that people love most: their cars and their wall sockets. Electricity is the relatively easier problem, because there are more options and fewer sources—in the United States there are about 500 power plants and several hundred million automobiles. A variety of technical options are available for both.

Rubin grouped policy options into carrots and sticks. Technology policy options were characterized as voluntary incentives (carrots), covering R&D and other popular programs such as voluntary recycling or thermostat resettings. Resolving the climate problem will also require thoughtful discussion of regulatory policy options (sticks) to drive actions that would otherwise take much longer or not occur. Policy recommendations from the NRC study included a mechanism for economywide carbon pricing coupled with other types of regulatory policies, new research centers, and heavy investment in R&D.

He applauded public and private sector initiatives (such as that of former President Clinton and New York City Mayor Michael Bloomberg) focusing on climate issues in cities, and recognized the European Union for its efforts on international targets, although he noted that these are mild considering what is needed to reach an 80 percent reduction in CO_2 by 2050.

Research has developed analytical solutions, and Rubin showed five computer models that found solutions to reach the 2050 target. While the solutions differed and involved a variety of technologies, all showed that major changes in the energy system would be needed. The proposed emission budget range for 50-80% reductions was technically possible but could be very difficult to achieve. It would also be costly, as all the models showed the GDP growth rate slowing to 0.5–2 percent. Developing advanced technologies or innovating more quickly than has been achieved historically could substantially lower the cost of mitigation.

Discussion

In response to audience questions, Rubin indicated that the kinds of new educational initiatives proposed in this project can make a difference. His program at CMU has developed an interdisciplinary course that brings together technical and policy students, graduate and undergraduate, to work together on solving a problem; the effort builds teamwork and interdisciplinary understanding in a project environment to which students are unaccustomed. Surveys well after graduation show great appreciation for this course.

He also reminded the audience that other problems that warrant attention include the lack of political consensus on some key issues; the role of developed versus developing nations; the costs of mitigation, the best way to do it, who might win and lose; the availability of options at a meaningful scale; and social acceptability (e.g., comparing nuclear to coal with carbon capture and sequestration or storage or other options). In fact, although the NRC report underscored the urgency of the issue, it is not being attended to and organizations and countries today are generally not really willing to do what's necessary, he said.

Engineering Perspectives: Toward Structural Change

Jackie Kepke, a consulting engineer working on public infrastructure projects and global technology leader of CH2M Hill's water management portfolio, looks at climate change risk assessment and adaptation planning. She spoke about public infrastructure designed for water and how to keep it functioning through challenges associated with climate change as well as urbanization, growing developing-country populations, and water scarcity, among others. These pressures compound each other and influence the management of natural resources and infrastructure.

CH2M Hill is a global company with clients of all scales that does big infrastructure projects in a sustainable and climate resilient way. The company does work for city governments, for private and public entities, and may be involved in responding to an energy challenge, constructing a building, or modifying a water system. It works to bring those parties together for integrated systems and solutions. It also educates its 25,000 employees about the challenges of climate change and works to inculcate these challenges in their thinking. And because its consulting engineers have an ethical obligation to think about them, the company takes its approach to clients who may not be asking for climate-resilient solutions.

Temperature increases, more extreme floods, droughts, and storms, rising sea levels, and ocean acidification all impact water infrastructure, the natural water system (source water, storm water, water treatment, waste water), and agriculture. What is the best way to consider all these components of water management in the context of climate change?

Unequal impacts of climate change present a further challenge. Geographic impacts, needs for community support, and questions of social justice require attention. Balancing agricultural, rural, and urban water needs will be difficult. In an increasingly urbanizing and drought-prone environment, how can resources be allocated fairly and ensure food security for growing city populations?

The Need for Systems Approaches

Kepke stressed the need to break down traditional engineering silos in favor of integration and systems thinking. Public infrastructure requires thinking about transit, waste, energy, water, and buildings as integrated systems, using tools that help balance resources to optimize city infrastructure. It is not adequate to build water infrastructure to handle whatever is sent down the sewer; it is better to engineer buildings to produce less sewage and reuse it and to pursue associated recycling strategies for the city as a whole.

Similarly, more informed public policy will require breaking down silos in the public policy arena. Many cities have a department of environment for energy and climate planning and a separate water department for water services and waste water. These different city services need to communicate with each other. Unfortunately, communication between cities or between city and state is even less common.

Water Management

Kepke explained the need to bring back the water cycle in water portfolio management. The water cycle is continuous and integrated: humans today are using the same water that the dinosaurs used, and drinking water, waste water, and storm water are not dissociated, as different engineering programs present them. The water cycle is a way to track different units of water through the system for optimized management; it requires thinking about recycling, rainwater capture, and system integration. The water cycle approach should be incorporated in engineering education and adopted by practicing engineers.

Kepke presented some examples of the scale of the challenges in water management. The lower Colorado water supply system, because of decreasing water availability, presents conflicts between agriculture and urban use and between the cities of San Antonio and Austin. Parties to the disputes have vastly different ideas about how the system should be managed and about the implications of climate change for the system. And California is trying to solve longstanding problems of declining smelt populations and other ecological challenges in the San Francisco Bay Delta system. The Bay Delta Conservation Plan uses probabilistic modeling to do climate projections to balance urban and agricultural water needs, ecological and human needs.

What are the best ways to bring together the needed scientific and public policy perspectives in order to address these complex problems, and, using appropriate communication strategies, provide recommendations to adjust the capital planning strategies of these cities and their water utilities so that they accommodate the myriad ways in which changes in climate and weather patterns affect water supplies and uses? How can science and engineering promote decision making that allows forward movement on such challenges, rather than further delay and yet another study to try to resolve disagreement?

Investment and Other Strategies

Infrastructure investment is critical. In 2009 CH2M Hill prepared a study to inform Congress on the investments needed from US waste water and water utilities to adapt to climate change by 2050 (the calculation did not include the $500 billion estimated shortfall for public water and waste water infrastructure investment).[11] The study sought to make the point that climate change legislation should consider not only how to mitigate climate change but also how to fund public infrastructure adaptation.

Utilities and other public and private entities are making huge investments to maintain and upgrade their infrastructure, Kepke observed, and may not believe that they have the time or money to figure out how climate change would change their plans. CH2M Hill tries to help its clients see that climate is one risk among many that they already deal with. Utilities are comfortable with the concepts of security risks, vulnerability assessments, and asset deterioration. Climate is another risk to manage in that adaptive context, so utilities must identify the hazards (e.g., resource management during more frequent severe storms, pump failure because equipment is too old or rusty), do the risk assessment, and manage the risk. Though it is difficult to think of these risks in the same context because the timelines are not the same, this is a sensible approach.

Kepke proposed an adaptive management strategy: Rather than laying out all the investments now, make sure that the monitoring and data are in place to identify initial impacts of climate change on the system and think about what investments to make at certain trigger points as changes occur. But with this type of incremental adaptation it is important to ask, At what point will the strategy no longer work? At what point will a strategy to make wet wells and pipes just a little bigger fail and will facilities need to move out of the floodplain? When does a paradigm shift happen?

Mitigation strategies are also needed. In cities water and waste water treatment is a significant source of greenhouse gases because of the energy demand associated with treating and pumping water.[12] A combination of mitigation and adaptation approaches is essential so that new infrastructure systems do not require even more energy; for instance, water recycling based on reverse osmosis is very effective for combating climate change because it is a resilient supply, but it uses a lot of energy. In some cases, rather than moving straight to desalination, fixing leaks in the system has adaptation benefits: not as much water needs to be pumped, and less water means less electricity for treatment.

---

[11] See http://www.ch2m.com/corporate/water/climate-risk-resilience/confront-climate-change.asp.
[12] http://www.epa.gov/climatechange/ghgemissions/sources.html.

Discussion

In the question and answer session, Kepke mentioned the importance of long-term savings to influence near-term public planning toward sustainability. Organizations spending $2 billion to put new pipes or green infrastructure in the ground want their work to last as long as possible under all conditions; so spending $20,000 to do an analysis of the projected future climate record, and a sensitivity analysis of how it differs from the historic climate record, is a worthwhile investment. To make such planning a priority the US Environmental Protection Agency may need to work with the states to encourage these investments.

Kepke called on the engineering education community to help engineers learn how to integrate policy contexts into engineering decisions because that is what they will face in practice. They will see that decisions are made not based simply on the best engineering solution but by bureaucracies and policymakers based on financial constraints, and that the best engineering solution *in that context* moves forward. Helping engineers understand and deal with this reality is important.

Asked how common her thinking is in the engineering community, Kepke said that large companies typically have many different kinds of expertise available in-house, to achieve both competitive advantage and public welfare.

For outreach and education, she suggested communicating with relevant professional societies such as the American Waterworks Association, the Water Environment Federation, and the International Water Association in the water management field. Greater academic input in their conferences would be valuable for consulting engineers who attend since their clients are there. Academics also have much to learn from practicing engineers who work in the United States and abroad.

Engineering, Engineering Education, and Climate Change

David Daniel, a geotechnical engineer serving as the fourth president of the University of Texas at Dallas, talked about the challenges climate change poses to engineering education.

Limitations of Data-Based Approaches

Engineers are trained to demand facts and data: if a salesperson tells engineers that a new reinforcing steel bar is stronger and more corrosion resistant than the old steel, their reaction is to demand test data, the long-term corrosion tests. This demand for evidence makes for safe bridges and reliable function. But the insistence on retrospective data to drive design is actually an important educational challenge, because climate change poses unprecedented conditions that do not have historical data.

Daniel described two categories of engineers who design structures in light of uncertainty. One uses a probabilistic analysis of past data and judgment about a design criterion. For example, they might review runoff data in a river over a period of time, draw a histogram, and make probabilistic decisions such as design for 100- or 1,000-year flow. Unfortunately, many people without technical training do not understand probability or the meaning of "100-year flood." So another approach is simply to look at the historic records. In New Orleans, for example, the levees were originally sized in the mid-20th century based on the worst Gulf hurricane of record at that time—which was not a 100-year hurricane. Similarly, all of the water design in Texas is based on the worst drought in the last 100 years, with no rationale except that legislators could understand it and so they wrote it into law.

Practicing engineers have a hard time, he said, embracing probabilistic designs and are much more comfortable with deterministic designs. For a building designed in a deterministic way, the strength of a piece of steel or concrete has a number: loads on the building from wind, people, and the weight of everything in it generate a number, and the engineer makes sure the building strength is greater than all the loads on it. That is a straightforward deterministic analysis.

Probabilistic analysis involves combining histograms of the strengths of the concrete, steel, wind, and other loads to yield a probability of the circumstances where the loads exceed the strengths and the building collapses. But this more rational way to design is harder to communicate; it is much easier for engineers to say that something *will* or *will not* break, as opposed to providing a probabilistic analysis and communicating it in a way that people can understand.

Lessons from New Orleans

New Orleans is of critical significance to engineers because rising sea level in heavily populated and industrial low-lying cities has such enormous and obvious impacts. In addition, the city will probably sink about a meter over the next half century or so and face increasing maximum-intensity hurricanes as the Gulf of Mexico warms.

The Corps of Engineers has talked about upgrading the New Orleans levees for the 100-year flood. As a point of comparison, the Netherlands designed its levees for the 10,000-year flood, but the difference there between the 100-year and 10,000-year storm surge is not that great as there is no huge range of differences in North Sea storms. In contrast, in the Gulf of Mexico the difference is enormous, so going from a 100-year to a 10,000-year design is far more difficult.

The challenges to New Orleans have never been dealt with very well, according to Daniel. The levees are designed at best for a 100-year storm; worse, they are designed not to be overtopped and in fact are built of materials that will self-destruct if overtopped. While it is only a question of time before the levees are overtopped, it is politically very difficult to expand them.

To prepare for climate change, an upgrade to the levees should provide for the likelihood of raising their heights. The design should be reasonably easy and economical. If money isn't available now to make them higher, engineering education and practice should recognize the probability in the future.

Designing for Change and Resiliency: Educational Opportunities

Engineers design for change all the time. For example, they design for a deteriorated material, such as a plastic pipe exposed to sunlight and degraded by UV radiation: they test for accelerated UV degradation and use retrospective information to project into the future.

But engineers have not been taught to think about designing for something in the future with no historical basis for that design. Even in homework problems engineering students work with specific numbers. Take sea level in the year 2100, for example; a homework problem on the topic would require a number or a range of numbers. Where does that number or range come from? Might a set of design criteria provide an educational tool? One of the great opportunities with education is the leeway to do almost anything as long as it is reasonable. Thus an instructor could ask students to pick sea level in a few cities, incorporate a median estimated curve and a range of numbers, pick the loads, and design the building for those loads.

In the absence of retrospective information, what ought to come out of this educational endeavor is a new way of thinking about changes in the future that cannot be predicted by anything that has happened in the past. Daniel gave the example of multiple megacities and the ways they might challenge the nation's infrastructure. Nothing that has happened before can help predict the results. This new way of thinking would be a lasting contribution to engineering education, helping engineers to be forward thinking in ways counterintuitive to the way that engineering usually gets done. It requires forecasting, which has scientific credibility and is different from standard engineering approaches, which rely on historical data and demonstrated capacities.

Fundamental challenges for climate change specialists are those of probability and risk, and these need to be conveyed even to skeptics of climate change in light of the associated risks. Daniel suggested using "resiliency" to label engineering for climate change. There is at the very least a risk that the climate change projections for sea level rise will turn out to be correct, so engineering should ensure resilient design and construction in the face of those projections.

In closing, Daniel urged consideration of global change in a quantitative way that is consistent with the way engineers are taught to think about data. Manufacturing provides business-driven opportunities that may be the most immediate recourse. Complex systems engineering for consideration of food chain issues, for instance, is imperative; the study of such issues would have spin-off benefits and relevance to a lot of other educational programs. He proposed a design goal for educational purposes; for example, an environmental engineering course might develop projections for sea level change that could be used in many kinds of courses.

Finally, he pointed to the need to build a network of people who come at the issues from different perspectives and can develop good lines of communication. With a focus on education, this network can encourage interdisciplinary communication and development of a common language, which is a great place to start.

Discussion

In the question and answer session, Daniel agreed on the urgent need to train engineers for the growth of the developing world; he predicted that private firms, not academia, will likely drive the response, with rapid development and enormous business opportunities in product sales for the rising middle class and global infrastructure needs. Certain industries understand that their business models need to include climate, such as the insurance, water, and earthquake design industries, which also have a history of using probabilistic analysis.

Professional societies will have a role in continuing education, given the limited hours available in the engineering curriculum. Different predictions of sea level rise pose a probabilistic problem that needs to be communicated well, he concluded, and expressed optimism about the ability of untenured and junior faculty to enter this field and gain tenure.

Geoengineering Potentials and Myths

Alan Robock, a distinguished professor of climatology in the Department of Environmental Sciences at Rutgers University, began by asking, What shall we do about climate change? Some say that mitigation can slow climate change, and doing it now will be cheaper than waiting to study the impacts and adapt. Explanations for some cycles include the role of volcanic eruptions and atmospheric pollution (and recovery from them). Some people have argued that if this happens naturally, why not do it on purpose?

But Robock stated that geoengineering, or trying to control the climate system, is not the answer. He noted two technically different approaches to geoengineering with different ethical and policy implications: carbon capture and storage removes gases that cause warming, and solar radiation management blocks out the sunlight. He focused on the latter.

Solar Radiation Management

There are four techniques for solar radiation management:

- an aerosol layer in the stratosphere that mimics the climate effects of volcanic eruptions,
- the seeding of clouds in the troposphere to make them brighter so they reflect more sunlight,
- surface brightening to reflect more sunlight, and
- space reflector satellites or a "cloud" of satellites [not discussed].

Of these, the two most feasible methods, which have also gotten the most attention, are cloud brightening and the stratosphere aerosol layer.

Ships have been designed to spray ocean salt up into the clouds to make them brighter so they reflect more sunlight. Several climate model simulations have examined how this might work in areas around the world with low clouds—off the west coast of North and South America and off the coast of Africa; they found a large reduction in precipitation response over the Amazon. Advocates have said that the brightening can be quickly stopped, but drought in the Amazon may be more difficult to stop.

The stratosphere plan has gotten the most attention. With diminished hopes for mitigation, Robock indicated that several scientific papers have stimulated this attention by postulating that, since volcanic eruptions cause cooling, emulating them could help. (More on this approach below.)

Technical Challenges

Robock does not believe any of these proposals stands up to careful scrutiny. Ideas for surface brightening do not seem particularly promising. Painting roads white doesn't last long and covers a small area. A model examining brightening leaves found that their surface would not get as warm, resulting in less evaporation and fewer clouds, so the effect would be warming and not cooling. Putting bubbles in the ocean to make it brighter has been proposed, but there are a number of technical difficulties.

One idea is to engineer particles that will not destroy ozone and place them appropriately in the stratosphere to achieve the desired effect. But stratospheric winds would blow them around and numerous negative consequences are likely—regional climate change (e.g., temperature and precipitation changes); rapid warming when the particle seeding stops; potential inability to stop the seeding rapidly, particularly in emergencies; continuing ocean acidification; ozone depletion and enhanced acid precipitation when the particles return to earth; whitening of the sky and less solar radiation for solar power; effects on plants of the diffused radiation; effects on cirrus clouds; and environmental impacts of aerosol injection.[13]

Using a state-of-the-art climate model from NASA, his research team modeled putting particles in the stratosphere over 20 years and then stopping for 20 years. Previous uses of the same model to simulate volcanic eruptions did very well, so this approach has some credibility. For precipitation changes the model found a reduction of precipitation over Africa, India, and China, where several billion people live (just as previous volcanic eruptions have resulted in lower river flows and less precipitation). And cooler continents and warmer oceans drive a weaker monsoon.

[13] Robock, Alan, Luke Oman, and Georgiy Stenchikov, 2008: Regional climate responses to geoengineering with tropical and Arctic SO2 injections. J. Geophys. Res., 113, D16101, doi:10.1029/2008JD010050. See also: Robock, Alan, Allison B. Marquardt, Ben Kravitz, and Georgiy Stenchikov, 2009: The benefits, risks, and costs of stratospheric geoengineering. Geophys. Res. Lett., 36, L19703, doi:10.1029/2009GL039209.

Data also show that after volcanic eruptions chemical reactions on the particles release chlorine, which destroys ozone. Another effect of volcanic eruption is less direct light because of a lingering thin cloud that scatters the light, which may interfere with solar energy production.

Difficult Questions and Moral Hazards

The results of these simulations raise difficult questions. If global average temperature could be controlled, what should it be? Should it be constant? Set at a 1980 or an 1880 average? Suppose Russia and Canada want it a bit warmer, but islands in the tropics want it cooler—who decides? There is no governance mechanism for such decisions. What if the project were discontinued and global warming progressed much more rapidly than if nothing had been done? An increased rate of change would be really dangerous, he warned.

Robock cited further concerns: the potential for human error, and recognition that much is unknown. There are moral hazards associated with belief in technological fixes. People might embrace the possibility of geoengineering and ignore needs for mitigation. Geoengineering might be used for military purposes (although a treaty might prohibit such use). Stratospheric aerosol would interfere with Earth-based astronomy and affect star gazing, satellite remote sensing, tropospheric chemistry, and passive solar heating. It would increase sunburn and effects on airplanes flying in an acid cloud up in the stratosphere .

On the other hand, Robock acknowledged reasons why geoengineering might be a good idea. It could cool the planet and reduce or reverse sea ice melting, ice sheet melting, and sea level rise. It could increase plant productivity, provide an increased sink of CO_2, and have unexpected benefits. Can these potential effects be quantified so that society can make an informed decision about the use of geoengineering in the future?

Robock reminded the audience that continuing with business as usual will continue to put out more greenhouse gases, which will force more climate change. Mitigation can reduce the forcing. Carbon capture and storage can also reduce it and potentially solve the problem. But suppose dangerous levels of climate change are reached before then? So far the only reasonable approach is solar radiation management to temporarily reduce the most dangerous aspects of climate change. But would this approach add to the dangers? That is where more research is needed, as set forth by the American Meteorological Society and American Geophysical Union (Box 3.1).

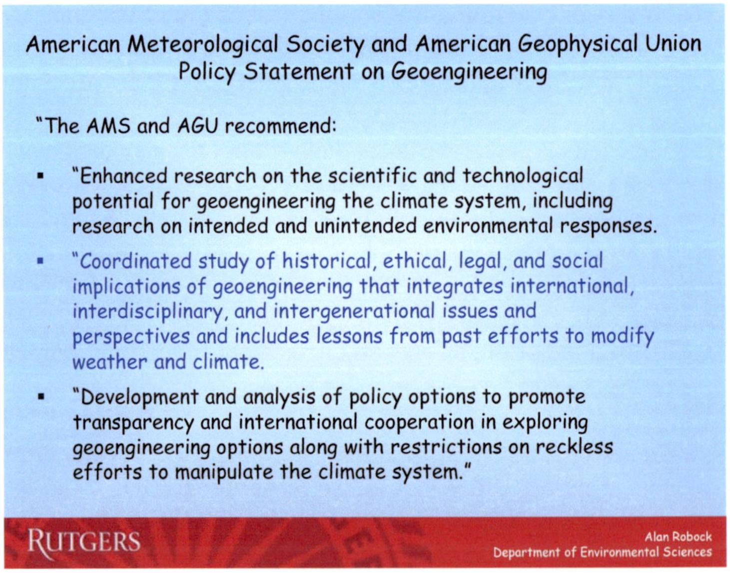

Box 3.1

The UN Framework Convention on Climate Change convinced the United States to prevent anthropogenic interference with the climate system; the United States thought that such interference referred to the production of greenhouse gases, but, based on the concerns described above, Robock thinks geoengineering should be added to the Framework pledge.

Discussion

In the question and answer period, Robock pointed out that what is dangerous is different for different individuals. Climate change already is bad for some people, and for others it will get worse and worse. The idea to return to 2°C above preindustrial levels, which was agreed on in Copenhagen, is arbitrary and is a distribution, not a number.

A statement issued by the American Meteorological Society and the American Geophysical Union defines geoengineering as intentional control of the climate system. Robock believes there is a distinction between intentionally controlling climate and doing it inadvertently. He pointed out that direct air capture, which involves removing CO_2 from the atmosphere through some chemical means and burying it underground, can be done. However, it is more expensive and less efficient than taking CO_2 out of the smokestacks of coal-fired power plants, and much energy is needed to regenerate the chemicals for each cycle of direct air capture. There is also the question of where and how to sequester the CO_2. Mechanisms to remove something that is causing harm would be a great thing for engineers to work on.

Robock is a member of the UK Royal Society's Solar Radiation Management Initiative, which has been discussing the ethical implications of geoengineering (which Robock notes could be viewed as intentional pollution in the name of science) in the atmosphere or in the ocean. A London Convention on pollution of the ocean stopped some experiments, but for much research, particularly in the ocean, there is no ethical guidance and there are no stipulations about what impacts to watch for. Within a country there may be regulations, but they do not apply to work at sea.

Remarks on the Presentations

Katie Johnson, an electrical engineer and director of the Center for Research and Education in Wind at CSM, saw four themes running through the session. One was the need for engineering education beyond the technical aspects. Social justice, for example, is both a domestic and an international issue, in terms of who "controls the thermostat." The second recurrent theme is the urgent need for transformational, not just incremental, innovation.

Third is the question of public support of funding for the necessary research, and the need for engineers to clearly convey what they can do. The decision to fund the space program to go to the moon is an example of designing for future events with little to no historical record, but the societal support was there to do it. That excitement may be needed today to drive the necessary transformational changes.

The fourth key point was that problems involving food or water or health do not necessarily look like they involve engineering to people who are not engineers. Engineering education needs to prepare students to recognize and address this lack of understanding.

David Slutzky, a research associate professor in the Department of Urban and Environmental Planning as well as the Science, Technology, and Society Department at the University of Virginia, agreed with Johnson and reiterated the theme of urgency: The timeline for improving the engineering curriculum is short, and incremental change will not suffice. In addition, the response needs to be broad and inclusive. Integration of climate change perspectives throughout traditional engineering education is necessary, as is integration of engineering with policymaking and business. Economic and social questions also have to be addressed, such as: Who pays and how? Who makes sacrifices?

Solutions will require integrating engineers' thought processes and actions with those of other disciplines. Perspectives from the field of science and technology studies (STS) are relevant here, but will have to be much more directed if they are going to help with engineering approaches to climate change. STS faculty members will have to teach engineering students how to be good ambassadors in this area.

Engineers also need to become comfortable with not just uncertainty but also unknowns, although it was not clear to Slutzky how that can be incorporated systematically in the engineering disciplines. Participants suggested a certification program and the use of case studies, which make discernible the necessary steps in a process to get things done. For instance, discussion of hurricane Katrina can illustrate the relationship between land use policy decisions about marshlands and the level of inundation that occurred in New Orleans, and highlight the political arguments about levee design.

What are the best ways to communicate about abstract—and sometimes hard to support—ideas about the projected negative consequences of not doing something, or of doing something poorly, to address a threat that is perceived as unproven? Slutzky posited that resolving this and other issues identified in this session would require involvement across multiple sectors—from undergraduate engineering education to the professional societies to practitioners and customer interfaces—to ensure that difficult choices receive adequate consideration.

General Discussion

In the few minutes remaining, participants noted that the need for innovation can conflict with the need for well-accepted standards for engineering practice, particularly in contexts where litigation is likely to be an issue. But careful research can promote innovation, and junior faculty in particular can be encouraged to undertake innovative projects.

Although improvements in weather forecasting enable better response to catastrophes, TV weather forecasters are just beginning to understand and come to grips with climate change in their reports. The American Meteorological Society committee that wrote the recommendations for undergraduate education recently added a requirement for the addition of climate change to the undergraduate meteorology curriculum to improve understanding among weather forecasters and other meteorologists.

Examining Interventions

In the capstone workshop, two speakers examined risk and adaptation strategies from engineering perspectives, a third presented a case study of adaptation to sea level rise, and the fourth focused on assisting climate decision making when facing scientific uncertainty.[14]

Armin Munevar, global water resources director, CH2M Hill, presented "An Engineering Perspective on Climate Adaptation, Risk, and Resiliency." He noted that climate change is leading firms to recognize that past, prediction-oriented approaches to reducing and managing uncertainty need to incorporate approaches that accept irreducible uncertainties and emphasize resiliency, robustness, and adaptive management, for instance through flexible design approaches. He used scenario planning for the Colorado River system as an example and underscored the need for partnerships among science/academia and the private, public, and NGO sectors to address the problems.

David Lapp, P.Eng., manager, professional practice, Engineers Canada, Secretariat, Public Infrastructure Engineering Vulnerability Committee (PIEVC), discussed "Infrastructure Climate Risk Assessment in Canada: An Engineering Strategy for Adaptation." Engineers Canada, the country's national body for the

[14] Video and slides from these capstone presentations are available at www.regonline.com/builder/site/tab2.aspx?EventID=1155563. The summaries here point out the range of technical and policy options underway or under consideration to address issues of climate and infrastructure needs.

profession of engineering, accredits Canadian undergraduate engineering programs, is a member of the World Federation of Engineering Organizations (WFEO), and chairs the WFEO Committee on Engineering and the Environment. Lapp described the protocol PIEVC has developed to examine the risks climate change poses to Canadian infrastructure and ways to mitigate and adapt to those risks.[15]

Greg Kiker, associate professor of agricultural and biological engineering at the University of Florida, looked at Florida's vulnerability to sea level rise. He described the association between the state's growth and development and sea level rise, and reiterated the difficulties that face decision making in complex, coupled human-natural systems. The use of numerous models, scenarios, and data sources is necessary to develop integrated information for use in decision making about managing habitat and development in Florida's most threatened coastal areas in the 21st century.

Using a case of storm surge, Robert Lempert, director, Frederick S. Pardee Center for Longer Range Global Policy and the Future Human Condition, RAND Corporation, spoke about "Informing Climate-Related Decisions When the Science Is Uncertain." He stressed the need to acknowledge unpredictability when dealing with the deep uncertainties of climate change and to face these challenges when making investment decisions. He recommended a backward analysis—looking at what happens if a strategy fails to meet its goals—and including stakeholder engagement as a way of moving ahead on infrastructure investment. He analyzed two cases, Port of Los Angeles infrastructure planning and the Louisiana Master Plan for a Sustainable Coast, to demonstrate the merits of this approach.

The workshop also included a panel on uses and plans for Colorado River water resources as they affect the southwest region of the United States. Carly Jerla, comanager of the Colorado River Basin Water Supply and Demand Study, US Bureau of Reclamation, reviewed the study results, including a variety of criteria for assessing options in the near and longer terms as well as next steps toward implementation.

Kay Brothers, former deputy general manager of the Southern Nevada Water Authority (SNWA), explained the effects of drought on the SNWA decision to construct an additional water intake lower in Lake Mead.

On behalf of Chuck Cullom, Central Arizona Project (CAP) geologist/hydrologist, Mohammed Mahmoud reviewed the history and characteristics of the CAP aqueduct, designed to divert the remainder of the Arizona allocation of the Colorado River for urban and agricultural use. A seven percent reduction in the Lower Colorado River Basin would result in a 30 percent reduction in CAP water allocations.. Planning and adaptation studies are under way.

Clifford Neal, water resources advisor for the city of Phoenix, reported that the city has adapted to growth by developing numerous programs to stabilize water demand and wastewater generation. Future shortages in supplies and impacts of climate change could require more interventions, such as expanded local well capacity, underground storage and recovery, demand management, fees, and river augmentation.

In Summary

Presentations and discussion indicate that a wide range of interventions to develop and use engineered systems in society are under discussion and underway. They are intended to address the influence of climate change on sociotechnical systems. The implications of these developments for engineering education and education more broadly may be a fruitful area for the CCEP to explore.

---

[15] For a more detailed account of Engineers Canada initiatives, see the section on Engineering Professional Societies in chapter 7.

Chapter 4
CROSS-CUTTING THEMES

The final panel session in the first workshop addressed the cross-cutting themes underlying the project focus: justice, sustainability, and governance, trust, and public engagement.[16] Speakers articulated the diverse elements that make the definitions of these terms complex and contested, and examined how these concepts relate to climate change, engineered systems, and society and need to be given priority in education about the subject. Highlights from workshop presentations at the capstone workshop that address these themes are available at the end of this chapter.[17] Deborah G. Johnson, Anne Shirley Carter Olsson Professor of Applied Ethics in the Science and Technology Studies Department at the University of Virginia, moderated this session.

Justice

Joseph DesJardins, associate provost and academic dean and professor of philosophy at the College of Saint Benedict/Saint John's University, considered historical and philosophical perspectives on justice, which may be helpful in the context of climate change impacts and decisions.

He began with John Rawls' definition of justice as "the first virtue of social institutions," involving basic liberties and fair access to primary goods (such as food, shelter). In this definition, justice or fairness can offer a modus vivendi for getting along when interests or ideals of the good life conflict. In contrast, in Book One of Plato's *Republic* Thrasymachus' view is that justice is nothing but the advantage of the stronger. DesJardins pointed to the siting of power lines as an example of Thrasymachus' approach to decision making. In a recent case in central Minnesota, an administrative law judge approved the siting of a line to avoid the university's campus and instead go through the farms of a less powerful group of constituents.

Thus, in considering decisions and political debates that affect social justice, the influence of political power needs to be recognized, although reason and rational public policy process should be the guiding principles in efforts to resolve questions of social justice.

<u>Distributive Justice</u>

Two modern theories illustrate differing views of what justice requires: libertarian theories, broadly defined here, define justice in terms of freedom from interference, DesJardins said, while egalitarian theories require consideration of the equal distribution of goods, particularly primary goods based on needs, not just the good of freedom.

Distributive justice concerns the allocation of benefits and burdens. There are going to be winners and losers; how should wins and losses be distributed? Plato, Aristotle, and Cicero taught that justice decides what each person is due and makes sure those benefits and burdens are distributed accordingly. But, DesJardins asked, What is to be distributed? According to what standard? And to whom—who counts?

Some classical Greek theories offer the following answer to the question of "what" is "due": there is a good life that humans ought to live, and the attainment of that good life constitutes justice. These theories

[16] Slides from the workshop presentations are available at www.nae.edu/21302/47874.aspx.
[17] Videos of these presentations and discussion are available at https://www.regonline.com/builder/site/tab2.aspx?EventID=1155563.

underlie many religious tenets. They contrast with more recent theories of justice that focus less on what people get than on how they are treated—disrespectful treatment is unjust treatment. Thus the shift was from *what* to *how*, from distributive to procedural justice.

In addition to these conceptions of justice, a third perspective has recently emerged in which justice requires the recognition of human capabilities to achieve wellbeing. Real freedom requires equality of capability to achieve well-being. Humans' innate capabilities—their lives, health, senses, emotions, imagination, relationships—make them deserving of the "primary goods" of welfare, income, liberty, and respect. Justice therefore requires that people be accorded the necessary conditions to achieve well-being.

Who Counts?

Turning to the question of "who counts" in the allocation of justice, DesJardins reported that, historically, the range was small: males, the wealthy, and citizens. Then the exclusions diminished and women and minority citizens were added. Now philosophical questions have expanded to include other groups and even living beings that are not human. Now when we say "to each his or her due," might "each" be an animal or other life form?

What about future generations? What is "due" to people who do not exist? DesJardins suggested that, in determining responsibilities to future generations, it is appropriate to consider justice for them. They should have the opportunity to live the same kind of lives as current generations—to pursue their own goals and meet their needs.

He invoked the Brundtland Commission, which defined sustainable development as "[meeting] the needs of the present without compromising the ability of future generations to meet their own."[18] He offered three specific ways to make the world "better than it would have been" for future generations: develop alternative energy, conserve resources, and limit population. He also cited the importance of preserving wilderness and animals for future generations.

Discussion

In the brief question and answer session, DesJardins noted that presenting justice in terms of a way to reconcile conflicting interests seems to work better with his business school students than more philosophically abstract approaches. Case studies also are effective. Audience members pointed out the importance of recognizing that artifacts or systems can enhance or curtail people's rights.

Sustainability

Paul Thompson, W.K. Kellogg Chair in Agricultural, Food and Community Ethics at Michigan State University, began by suggesting that sustainability is an ideal in the same sense that justice might be an ideal, and that discussions of sustainability today resemble those about democracy when Thomas Jefferson and John Adams served under George Washington. The two men had very different ideas of what democracy was about, but both were committed to it and to working within a framework; they argued at length about what the framework and democracy should contain.

---

[18] World Commission on Environment and Development (WCED) (chaired by Gro Harlem Brundtland). 1987. *Our Common Future*, chapter 2. Oxford University Press.

Historical Perspectives

Thompson reviewed the history of American thought about sustainability since the early 20th century and Liberty Hyde Bailey, the most famous agricultural scientist of his time and founding dean of Cornell's College of Agriculture. Bailey was worried about the social sustainability of agriculture as he saw farms disappearing and circumstances undermining the stability of rural communities. Aldo Leopold was similarly concerned about game management, averring that "a thing is right when it tends to preserve the integrity, stability, and beauty of the biotic community. It is wrong when it tends otherwise."[19]

With the beginnings of an environmental movement in the 1960s (following the publication of Rachel Carson's *Silent Spring* in 1962), sustainable agriculture began to be talked about in the 1970s. When Thompson started working in agriculture studies in 1980, there was a fairly robust debate about what sustainability might mean. That debate was overtaken in 1987 by publication of *Our Common Future*, by the Brundtland Commission. The definition of sustainability in that report still strongly influences the way people think about the subject, although it focused on sustainable *development.*

The Brundtland Commission report precipitated a decade or more of debate about what sustainability could be, winding up with what Thompson called "three circle" or "three P" sustainability: people (social justice), planet (functional integrity of the environment), and profits (resource sufficiency). He concentrated on functional integrity and resource sufficiency, but the discussion (summarized below) extended to the more recently recognized dimension of social justice (social movement).

Resource Sufficiency, Functional Integrity, and Social Movement

One way to think about sustainability is to consider a practice or system sustainable so long as the resources needed to sustain it are foreseeably available—that is, there is resource sufficiency. This way of thinking requires complex accounting methods to determine what is available, how quickly it is being used up, and whether there are replacements or ways to use less. Thus people who think about resource deficiency see sustainability as primarily an accounting problem, and appropriate policies or innovation practices follow from minimizing or overcoming the deficiencies.

A second way of thinking might be called functional integrity. A system or practice is sustainable if it is relatively invulnerable to the threat of internal collapse (i.e., something in its design or functioning that leads to its undoing). This way of thinking came primarily from ecologists but can be applied much more broadly.

Thompson illustrated with Leopold's approach to game management. In terms of resource sufficiency, wildlife populations are sustainable if there are enough animals to shoot. In terms of functional integrity, wild game populations are sustainable if the ecosystem that regulates habitat and population levels is intact (that is, there is a balance between predator and prey).

To put these ideas in an engineering context, Thompson said that resource sufficiency considers whether the system is efficient, whereas functional integrity asks whether it is going to break or not. To take an example relevant to climate, biomass electrical generation might be deemed sustainable as long as it complies with regulations and there is enough wood waste (or other biomass) for future use (that is, there is resource sufficiency). In terms of functional integrity, biomass plants are sustainable if they do not threaten processes that stabilize climate and air quality or regeneration.

[19] Aldo Leopold. 1949. *A Sand County Almanac*. Oxford University Press. p. 262.

The social movement aspect of sustainability relates to key goals and concerns—such as social injustice, imbalance of power, and lack of recognition—that are not accounted for in strictly economic or ecological approaches. A system or practice is unsustainable if it is generating acts of resistance, protest, and political change. This definition of unsustainability links the social movement dimension to the other dimensions.

Integrating Perspectives

Thompson pointed out that these different ways of thinking about sustainability rely on attention to systemic interactions, and discussed how scientific methods could incorporate the three approaches.

In terms of resource sufficiency, he posited that technology is good when it increases the efficiency of a production process or a consumption activity, or when it substitutes plentiful for scarce resources. It is bad when it increases total resource consumption. The "ethical maxim" is to find the optimal ratio between benefit and risk.

In terms of functional integrity, technology should not introduce fragility or brittleness into the system, nor should it create new sources of vulnerability. As a guideline he suggested a "precautionary approach": lack of full scientific certainty about risks should not preclude taking precautionary measures.

Last, in the context of social justice, technology is good when it levels power relationships, bad when it strengthens or entrenches them.

The various perspectives do not compete with each other in quite the sense that Jefferson and Adams had competing notions of democracy, so much as they indicate areas of emphasis in people's concepts of sustainability. Oftentimes, getting these values out on the table can lead to productive and complementary discussions about sustainability.

Discussion

Participants asked about the social movement perspective. Thompson explained that people involved in promoting sustainability seem motivated by social justice and even conflate the two. Why, then, use the term "sustainability"? To raise associated concerns about poor people getting pushed off their land in Africa, for example, by the rising costs of land and food and the push to develop biofuels. Such outcomes are not sustainable. Thompson agreed that it is important to talk about matters of social justice, as long as they are framed in terms of sustainability—that is, how they might lead to a social or political collapse.

Challenged to pursue this further, Thompson said he thinks the social movement perspective is problematic in two ways. On the one hand, it is not clear what the term "sustainability" adds to the discussion beyond that of "social justice". On the other hand, people who tend to define sustainability in social terms do not necessarily ask critical questions about other aspects of sustainability. For example, when social issues arose in agricultural discussions, those involved did not examine problems of sustained yield but thought instead about the fact that rural communities were getting poorer, farms were going broke—a sort of "de-development" was occurring. Rather than debating sustained yield versus development, Thompson suggested that functional integrity versus resource sufficiency is a productive way to think about sustainability and that a rich, deep discussion of the subject requires *both* perspectives, in addition to consideration of social sustainability.

Governance, Trust, and Public Engagement

Susanne Moser, director and principal researcher of the Susanne Moser Research and Consulting Company and a social science research fellow at Stanford University Woods Institute for the Environment, began with an engineering metaphor of Archimedes' lever and fulcrum to put governance in context. The levers, the determinants of adaptive or mitigative capacity in response to climate change, are economic resources, technology, information, skills, infrastructure, institutions, social capital, and equitable access to all of these. The fulcrum, governance, is the decisions, actors, processes, institutional structures, and mechanisms involved in determining a course of action. Governance is more than government; it may involve government actors but also others such as market mechanisms and civic actors. Because there is always tension between governmental and other actors, decision makers play a central role.

It is useful to look at decisions and decision makers in new arenas, where policies, laws, regulations, and related structures and institutions are not in place—and breakthroughs can occur through individual action. The context of climate adaptation provides just such opportunities. It also demonstrates the challenges to engagement and action, such as persistent and growing gaps between rich and poor in every country, and significant societal vulnerability to climate extremes even in developed countries.

Research on social capacity to respond to climate change indicates many barriers to progress, ranging from lack of leadership and policy guidance to ignorance; problems of coordination, collaboration, and communication at any level of government; lack of information; lack of funding (a problem that is becoming more pervasive); competing priorities; and legal obstacles to change and innovation. These difficulties arise in every sector and at every level, and are compounded by a sense of urgency and pressure for action.

Governance

Governance is the art and skill of turning capacity into action and, when the capacity is not there, generating it. It requires dealing with institutional issues, managing the political calculus, and making leadership more effective. Leadership and organizational culture are crucial in getting people moving.

In many instances regional cooperation is necessary—for example, in water resource management and coastal management—but the mechanisms are not yet established. And efforts at governmental devolution (sometimes known as "new public management") magnify the problem with unfunded mandates—asking more of local governments without providing the resources needed for implementation. Moreover, many US counties and communities no longer have planners, so the people with long-term perspectives who would be able to take action are simply not there.

Moser went on to observe that people who take action at the local or state level are usually *not* those with the best scientific knowledge or assessment. Rather, such leaders are creative and skillful in pulling together the necessary financial resources, and are persuasive or aware and strategic in handling social acceptability issues. She therefore believes that, as far as mitigation and adaptation are concerned, America is going backward right now. In her opinion adaptation plans are being developed in the absence of sufficient scientific information, and it would be very difficult to produce such knowledge fast enough to prevent some major maladaptations—efforts that will be insufficient, or have negative side effects or high costs, or even foreclose future options. In addition, constraints and failings in the current system may leave the country in a precarious position when catastrophes strike.

In light of all these concerns, the need for effective governance systems is only growing.

Trust

The perception of leadership is linked to trust, which Moser introduced with the example of a phone call she received asking whether she has diabetes. She put the question to the audience: If you were asked this question, would you answer it? The answer will depend on who's asking—a telemarketer, or an insurance agent, or your doctor, or your best friend—and the level of trust in that relationship.

Trust is not a fixed entity; it depends on context (geographical, social, temporal, political, economic, cultural), what is being requested, the perceiver's mental state (e.g., stable vs. paranoid), and past experience. Trust also depends on the scope and scale of the demand—or of the technology at hand. Questions of trust about, say, geoengineering are different from those associated with an immediate personal concern.

On climate change, studies show that Americans do not strongly trust any particular source of information, but somewhat trust numerous bodies, including federal and private scientific institutions. Opinions divide into six groups ranging from those most convinced and alarmed about climate change and most trustful of scientists and information sources, to those most doubtful. Trust in mainstream news media varies across groups but is lowest among the sources respondents were asked to rate.

For managing risk and communicating about it, trust is very much determined by perceptions of the communicators' knowledge, openness, and care. These moral aspects may be more important than expertise. People assess trustworthiness based on commitment, follow-through on a "promise" (implicit or explicit), and whether people are forthcoming with risk information. In social interactions, people tend to trust those who are like them, with whom they have deep connections and frequent interaction. These things are important to keep in mind in efforts to create or foster trust between people who do not have it. More trust is needed the less communication and perceptions rely on well-established commonly accepted science, creating a potentially very difficult situation with future climate management decisions.

Public Engagement

Why engage the public? Moser listed three principled reasons:

- Governments can't do it alone: achievement of major policy outcomes requires greater engagement and participation from citizens.
- Governments shouldn't do it alone: there are strong moral and political arguments for protecting and enhancing personal responsibility.
- Cost savings in doing it together: involving the public in active implementation/behavior change can be significantly more cost-effective than traditional service delivery.

She then reviewed some approaches to risk communication and public engagement. The first was to have experts tell people what they should believe and do. When that didn't work persuasion was tried, but that didn't seem to work either. Now the paradigm is to develop public engagement through two-way dialogue, which has the following benefits: it can build mutual trust, stakeholders understand risk assessment and response options, and decision makers learn and take into account stakeholder concerns.

Climate change engagement may differ from engagement in other areas because of the skepticism and uncertainty so often put forward, and because in-depth knowledge on mitigation is lacking and adaptation is still unfamiliar. Educational challenges are greater than on other issues, and audience interest and readiness are not at all clear. Persuasion about the need for mitigation and adaptation is needed, as are ongoing monitoring, learning, and repeated engagement. Local and state government are limited in their capacity but must be brought into the conversations.

True two- or many-way dialogue is rare, and it is not clear which segments of the public should be addressed, although better matches are needed between those who communicate and those who will be mobilized or engaged, she said. Most plans for adaptation, mitigation, communication, and engagement need to be strategically developed, recognizing that different people may mobilize or be mobilized for different purposes at different times in a dynamic process.

The Way Forward

Moser concluded with a summary of the reasons public engagement is needed:

- Mass/one-way communication is not enough.
- Dialogue can help to transcend impasse on deeply polarized matters.
- Change requires social support.

But the need for forums for deeper social engagement, ongoing dialogue, and support and accountability go unmet. In addition, there remain lingering questions: What mechanisms are needed to link across fields? What is the best way to effectively represent views about geoengineering at a global level, where such decisions will be made? If pressed to choose speed or thoroughness of engagement in an emergency situation, which will work?

She articulated the following steps to prepare the way forward (those that mesh with the focus of this project are marked with an asterisk):

- Rapidly and substantially expand multidisciplinary climate change research and development.
- Build technical capacity in *all* sciences (especially social science) and among decision makers.*
- Expand the nation's decision support capabilities.*
- Identify ways to provide financial and technical resources to governing institutions.
- Seriously engage the American public in the development and debate of a comprehensive climate risk management strategy.*

Discussion

In response to an observation about the difficulty of developing public confidence, given stakeholder fatigue, Moser explained that for people to remain engaged on an issue they have to view their participation as making a difference. She pointed out that social media might be very good for social mobilization around issues that can rally people quickly, but the level of engagement needed for an adaptation decision about where to site a windmill, for instance, would be different.

One participant asked about opportunities for engineers to engage with publics directly, rather than through local, state, or federal government institutions. Moser responded that although some governmental structures make public engagement difficult, there are often possibilities and even requirements for it. And engineers who work as consultants with any level of government or type of business are often involved in processes that engage the public as well as long as they adhere to procedural guidelines. In fact, more and more engineers and consultants are becoming involved in communication training because they see the importance of good communication. If staff are not trained to do effective engagement, a social scientist consultant can help.

The challenge is in bringing the parties together; facilitated dialogue and sophisticated help are needed to make constructive interaction happen. Moser said that the local officials she works with are often frightened of dealing with emotions, but she believes they can be a good place to begin a conversation. Communication that has nothing to do with conveying or persuading can make a connection by listening to and affirming the person, who is then more likely to be open to considering various points of view. This facilitated approach to dialogue helps people to be more comfortable with emotions instead of throwing stones at each other about building a windmill or a seawall.

Establishing trust is a psychological challenge more than an institutional one. People in local government or in local communities live and go to church with each other, their kids are in the same class, and those relationships prevail.

Climate Change and America's Infrastructure: Engineering, Social, and Policy Challenges

The capstone workshop revisited the project themes of governance, sustainability, trust and engagement, and social justice.[20] Two panels addressed policy and governance issues and justice and human rights issues in the context of climate change and engineered systems. In the first panel, speakers focused on the vulnerabilities for which government action and responsiveness are required and illustrated challenges and opportunities; the second set of speakers delineated ethical challenges and educational and community programs that can identify problems and propose solutions. The summaries below highlight material that relates most substantially to the project cross-cutting themes, adding to the views expressed above. For readers interested in learning more about what particular speakers had to say about a given subject, the complete presentations are available in the online posting of the video proceedings.

Policy and Governance Challenges and Strategies

Elisabeth Graffy, professor of practice and senior sustainability scientist at Arizona State University, introduced the session. She first stressed the need to figure out what to do when standard practices have not been working, in the face of social as well as environmental and physical challenges. Scientific findings can contribute to governance and management but do not provide answers. Important decisions must take account of political, social, and scientific differences. Systematic and generalizable knowledge about governance and management options, from researchers and practitioners, is also needed.

In addition, there is widely recognized value in identifying and examining community needs for infrastructure in efforts to address questions of climate, engineered systems, and society. And the notion of infrastructure can include engineered, social, and green infrastructures.

The three speakers in the panel provided specific examples of how localities and communities have responded and can respond creatively to the challenges that infrastructure needs pose to them.

The first speaker, Kristin Baja, hazard mitigation planner for the City of Baltimore Office of Sustainability, represented a city government perspective. She reviewed Baltimore's Disaster Preparedness and Planning Project (DP3), which incorporates climate change challenges in its planning and involves both hazard mitigation and climate adaptation. Climate projections indicate significant increases in average annual temperatures and precipitation, sea level rise, and increased flooding and storms, all of which affect human health. The Federal Emergency Management Agency (FEMA) and its Maryland partner as well as the Maryland Department of Natural Resources signed off on the city's plan, which was developed by the Office of Sustainability in a broadly participatory process.

[20] Video and slides from the capstone presentations are available at https://www.regonline.com/builder/site/tab2.aspx?EventID=1155563.

In the next presentation, on "Tribal Governments, Climate Change, and Infrastructure," Patricia Mariella, director of the American Indian Policy Institute, Arizona State University, talked about the need for advance planning and risk communication. The American Indian Policy Institute is tribally driven and integrative, and supports the community through, for example, a certificate program in tribal financial management. Mariella reminded the audience that tribes are governments and must respond as such. In Arizona, where much land is tribal and includes many mineral and other natural resources, tribes have water rights as well as opportunities in alternative energy, innovation, and regulation. Although much improvement is needed—for rivers, roads, and airsheds, for example—it can be achieved through multijurisdictional interaction. She finished by reporting what has been learned about communicating risks—and addressing fear and trust—with empathy and appropriately constructed messages.

Jennie C. Stephens, associate professor of environmental science and policy in the Department of International Development, Community and Environment (IDCE) at Clark University, addressed the third topic, "Post-Sandy Discourse on Energy System Vulnerability and Smart Grid," looking at the role of public discourse in creating conceptual connections between hazards, vulnerabilities, and climate change. The connections recognize human reliance on systems and infrastructures and the relevance of both mitigation and adaptation to that dependence. Her research results indicate that societal expectations (e.g., of reliable electricity), together with opportunities for technological upgrades and concerns about affordability, play an increasing role in public discourse. Government decision making should take account of the differences in perspectives and priorities revealed in such discourse.

Engineering, Justice, and Human Rights

Rachelle Hollander, director of the Center for Engineering, Ethics, and Society at the National Academy of Engineering, moderated this session and introduced the speakers.

Byron Newberry, professor of mechanical engineering at Baylor University, opened with a presentation on engineering challenges, in which he observed that attending to the priorities of people affected by catastrophe is a good place to start considering the challenges. Engineers need to recognize that people have a love-hate relationship with technological fixes, that populations include haves and have-nots, and that there will be difficulties in integrating diverse social aims and long-term risks in proposed technological solutions. He noted a number of pitfalls ranging from unanticipated failure modes and problems of physical or organizational interfaces and transitions, to poor communication, competing interests, and historical contingencies.

Barbara Rose Johnston, senior research fellow, Center for Political Ecology, University of California, Santa Cruz, focused her talk on "Climate Change, Human Rights, Justice." She used water to demonstrate the multiple values, physical and social, of a resource and how threats to that resource can interfere with social and physical well-being. Resolution of the threats may benefit some, while displacing and impoverishing others. In her view, sustainability requires biocultural health. Development of effective solutions requires acknowledgement of indigenous peoples as rightsholders who must be involved from the earliest stages of project planning, share in the benefits, and have the power to say no. Such involvement often results in better stewardship of resources, although it may not result in the greatest economic gains.

In Summary

Presentations and discussion acknowledged that effective response to the challenges of climate change on infrastructure necessitate attention to issues of sustainability, justice, public trust and engagement, and governance. Developing effective responses also requires formal and informal educational activities to address those issues.

Chapter 5
FORMAL EDUCATION INTERVENTIONS ON CLIMATE, ENGINEERED SYSTEMS, AND SOCIETY

This chapter draws on presentations at the second and final workshops that described overarching considerations as well as specific examples of effective interventions in engineering education on the subject of climate change.

Challenges of Incorporating Climate Change in Engineering Education

In a session on engineering education at the capstone workshop, Helene Hilger, associate professor emerita of civil and environmental engineering at the University of North Carolina at Charlotte, talked about challenges, current programs, and strategies for including climate change in engineering education.[21]

A key challenge is the rate of climate change, which is much faster than the rate of innovation in university engineering courses and curriculum. Furthermore, when new courses are offered, they are most often based on a faculty member's research or particular passion. As a result, classes are created on a case-by-case basis on topics that are not a regular part of the curriculum. But the alternative, from conception of a new course to its addition by the class registrar, can take a couple of years. There are no supports in the faculty reward system for the creation of new courses. Although teaching is considered in faculty tenure review at many academic institutions, research is often more highly rewarded, and faculty can place themselves at a disadvantage if they take the time and effort to develop new courses.

What's needed to compensate for these impediments are faculty who are not resistant to the topic of climate change and who are lifelong learners who can create course content from scratch. They also need to be brave enough to be associated with the sometimes controversial topic of climate change, humble enough to work with interdisciplinary faculty on the course, resourceful at gathering new materials on the topic, and generous with their time because they will not get much credit for this work.

Fortunately, a number of educational initiatives are beginning to address the incorporation of climate change in engineering. Among these are the Center for Sustainable Engineering, an NSF- and EPA-sponsored partnership of five universities that offers workshops and web resources for engineering educators (www.csengin.org), and a few university and professional society programs and classes (Box 5.1).

Hilger concluded with some ideas for the project team, colleges and universities, and sponsors to support efforts on climate change education in engineering:

- Create a recognition or certification program for faculty who are early adopters, to be considered in their tenure review.
- Engage the core engineering organizations: professional societies, professional licensing bodies, and accreditation bodies.
- Educate administrators on the importance of climate change engineering education so they can support and recognize their faculty.

---

[21] The agenda and video and slides of speakers' presentations are available at https://www.regonline.com/builder/site/tab2.aspx?EventID=1155563.

- Provide resource support to instructors teaching this material.
- Link grant funding on climate change education to multiple educational institutions within a state.
- Create a website for climate change educators, such as Stanford University's "Tomorrow's Professor" website (www.stanford.edu/dept/CTL/Tomprof).
- Create opportunities for students to see companies and employers acting on the importance of climate change and engineering.

BOX 5.1 Examples of Climate Change and Engineering Education Programs

Johns Hopkins University
Series of professional nonthesis degrees:
- Master of environmental engineering
- MS in environmental engineering and science
- MS in environmental planning and management
- Advanced certificate for post-MS study in climate change, energy, and environmental sustainability

Course on climate change and global environmental sustainability: multidisciplinary; critical assessment of science, impacts, mitigation, adaptation, and policy relevant to climate change and global environmental sustainability

Stanford University
Civil and environmental engineering undergraduate subprogram in atmosphere/energy with nuanced reference to climate change

Institute for Sustainable Infrastructure
Courses for evaluators and verifiers of the institute's Envision Sustainable Infrastructure Rating System

University of Michigan
Atmospheric, Oceanic and Space Sciences Engineering Department has an undergraduate concentration in Climate Impact Engineering, including courses on Earth's changing climate, core Earth system science, and environmental impacts on Earth systems

University of Montana
Joint program with Colleges of Forestry, Arts and Sciences, and Technology offers an interdisciplinary minor in Climate Change Studies that is open to all majors; educates students in three areas of climate change: science, society, and solutions

American Society of Civil Engineering
Certification in sustainability-themed courses, such as Fundamentals of Sustainability Engineering

Standards and Assessment of Educational Interventions

In a session at the second workshop, Richard Duschl, Waterbury Chair professor of secondary education, College of Education, Pennsylvania State University, reviewed developments and reports produced over the last 10–12 years in the learning sciences, primarily for the K–12 curriculum, that provide useful background information for pedagogical assessment.[22] He also outlined work being done to create the Next Generation Science Standards, for which he cochairs the Earth and space science part of the standards. Some of the disciplinary core ideas from the Next Generation Science Standards, specifically those for Earth and space science, are compatible with the concepts that the project team has discussed students should learn, such as the idea of Earth systems and the connection between Earth and human activity.

Duschl echoed the call to align goals, or what he referred to as standards, with the assessment of learning outcomes. Over the past decade much has been learned about learning, and measures and assessment techniques have become more sophisticated, so tools are now available to assess the knowledge and practices that are the goal of education.

---

[22] Richard Duschl also produced a paper on this topic for the project: "STEM Learning in Context: Opportunities and Challenges from Climate Science and Engineering, 2011, http://www.onlineethics.org/File.aspx?id=28160.

One approach that might be appropriate for the project is *learning progression*, which involves curriculum that fosters learning over a longer period of time and builds progressively on knowledge presented in previous courses. The project team might also consider both societal expectations of what students should know and the knowledge that students bring into the classroom. Asked to prioritize where the project team could focus its learning science work, Duschl endorsed the project's attention to problem-based approaches and what preparation employers want.

A helpful approach to identify learning goals is to ask, What would you like a student's enduring understanding of the course to be (i.e., after four or more years)? In the question and answer period, project co-PI Clark Miller, associate professor of science policy and political science, ASU, commented that if the goal was to "prepare engineering and public administration students who are going to be ready to enter the marketplace" and take leadership roles early in their careers to "help push forward thinking about and tackling the challenges that climate change poses to engineering infrastructure," then, based on Duschl's presentation, faculty need to think about developing the learning progression for these knowledge and skill sets.

Effective Interventions (1): Engaging Students for Ethical Action on Climate

Donna Riley, associate professor of engineering at Smith College, made the case that efforts to include climate change, its impacts on engineered systems, and the cross cutting themes in engineering education are "interventions" in the sense that they are out of the ordinary and may even conflict with what's in the textbooks.

She described the challenges she encountered when she added climate education to a standard engineering course on thermodynamics. Some students complained that the climate content along with the policy and ethics discussions about climate were not what they thought the course was supposed to be about and that these components should not be included in a required course. Motivating the students was crucial—they needed to understand why climate change matters, why it belongs in a thermodynamics class, and why reading, writing, and ethics are important to their education.

But teaching the topic helps students understand that it is naïve to believe that a simple approach, such as recycling or turning off lights, will suffice to address the problem and makes them aware that some people deny that climate change is occurring. Students also need to see faculty supporting this education.

To encourage students and practicing engineers to appreciate the value of good communication skills, Riley created an educational intervention based on Jim Hansen's book, *Storms of My Grandchildren*. The author describes his failed efforts in the 1990s to communicate about climate change to the US Congress. He learned that he did not have a clear, succinct story, that his communication was untactful, confusing rather than illuminating, and as a result provoked strong reactions against him.

In a second intervention, Riley introduced a semester-long Climate Action Project designed to (1) get students thinking on the big scale, at the society level, and at the level of engineering detail; (2) connect theory to practice; (3) connect their role as student to that of citizen; and (4) connect the college world to the "real world." Students were asked to determine how to significantly reduce carbon dioxide levels, in which "significant" meant 1,000 Tg CO_2 equivalent per year (14 percent of US 2010 output, about 1990 levels). Possible student approaches to the project included a small-scale demonstration or analysis of the sorts of structural changes needed to achieve the reduction. Students had to justify their method from a quantitative perspective, qualitative perspective (e.g., feasibility and effectiveness), and ethics perspective. Riley reported that the project had a major impact for some of the students and resulted in personal life changes.

A third educational intervention involved a case study essay on global climate agreements and the challenges of upholding them. This more traditional approach connected with current events and revealed injustices between the global North and South. Students analyzed the ethics (issues such as justice, governance, trust, and public engagement) of a climate agreement, either implemented or unsuccessful, from a variety of philosophical standpoints and stakeholder perspectives.

The final intervention used energy disaster case studies, such as nuclear energy accidents and the Gulf oil spills, to reveal problems with the governance and regulation of energy systems, structural inequalities in decisions about who pays for the costs (not only financial but also environmental and social), questions about responsibility for preventing and responding to such disasters, and the difficulties of designing for "unanticipatable" events. The cases demonstrate that disasters are considered simply "business as usual," and they raise questions about the feasibility of using nuclear energy to "bail out" of the climate change problem.

Riley concluded by reporting that as a result of her interventions her students enhanced their communication skills, improved critical thinking abilities, developed moral reasoning skills, became more socially engaged, developed some limited community organizing skills, and learned that nontechnical knowledge can complement technical engineering knowledge.

Effective Interventions in Undergraduate Engineering Education

At the second CCEP workshop, organized to examine the educational needs of different audiences from various perspectives,[23] speakers described effective interventions in undergraduate engineering education, particularly innovations that can both improve integration of climate change and engineered systems (CC&ES) in engineering curricula and scale up across institutions. In the first hour of the session invited speakers addressed specific questions that they had received before the meeting:

1) What are the unique challenges and opportunities to integrate CC&ES into engineering curricula?
2) What are the strengths and limitations of attempted innovations to bring new content into the engineering curriculum? Such innovations include
    a) Case studies
    b) Course modules team-taught by engineering and liberal arts faculty
    c) New courses on the particular subject, which are often treated as electives
    d) Workshops to prepare engineering faculty to develop and implement their own innovations (e.g., rewrite thermodynamics problems to include climate change)
    e) Online repositories (with case studies, readings, problem sets, etc.) that faculty can consult to bring new content into their courses
    f) Internally and externally funded grants to faculty to innovate
3) Are there other specific innovations that could be more effective, particularly to encourage faculty from different institutions to adopt them?

In the second hour, four project team members joined the speakers for a panel dialogue to further explore the topics. Audience members submitted questions to a moderator who presented them to the panelists for response.

Project team members Juan Lucena and Jason Delborne of the Colorado School of Mines (CSM) were responsible, respectively, for organizing and moderating the session.

[23] The agenda, video, and slides from the workshop presentations are available at www.nae.edu/Projects/CEES/57196/35146/62343/52752.aspx.

A Systems Approach to Educational Interventions

Ann McKenna, chair and professor in the Department of Engineering at the College of Technology and Innovation, Arizona State University, began by emphasizing the importance of aligning the proposed educational approaches (and the reasons for choosing them) with the goal of the project to integrate climate change in the engineering curriculum and with the values of the engineering community, and of assessing that alignment in the evaluation of those approaches.

More broadly, she urged the project team to take an approach that addresses the whole system of engineering education: the institutional structure and core working processes of teaching and learning need to be considered if engineering education is to be truly transformed.

- Changes in institutional pedagogy will require changes in faculty members' epistemological beliefs about how students learn.
- Institutions must be engaged to support changes in the classroom.
- Pedagogical products will need to be actively diffused (not passively posted on websites).
- Faculty and teachers will need to clearly see both the relative advantage to using them and how to incorporate them in the curriculum.
- Research is needed to identify barriers to changes in the curriculum.

Engineering educators will be key in these efforts to transform education, and can help think through what would be appropriate content and entry points for proposed interventions.

She concluded by encouraging the project team to network with groups that are also seeking to accomplish transformations in education. To that end she cited two 2011 meetings on transforming education. The first was a forum, sponsored by the NAE's Center for the Advancement of Scholarship on Engineering Education (CASEE), on the impact and diffusion of transformative engineering education innovations.[24] The second, a Purdue University meeting titled "Transforming Education: From Innovation to Implementation," was more broadly focused than engineering education.[25]

Tribal College Collaborations

Bob Madsen, professor at Chief Dull Knife College, a tribal community college in Montana, described connections that tribal colleges have made in science and engineering as an example for the CCEP team in its efforts to create effective partnerships with these colleges that could bringing CC&ES education into the tribal colleges.

Collaborations that have been the most engaging and productive for education have focused on research projects that resonate with the tribal community, such as the Engineering Research Center on water systems, involving Stanford, CSM, the University of California at Berkeley, and New Mexico State University. He also mentioned two NASA-supported collaborations: an engineering working group that involves 11 tribal colleges to establish preengineering and engineering programs, and tribal college

[24] National Academy of Engineering Forum on Impact and Diffusion of Transformative Engineering Education Innovations, February 7–8, 2011; agenda, papers, presentations, and associated materials are available at www.nae.edu/Activities/Projects/CASEE/26338/26183/26293.aspx.

[25] Purdue University Conference on Transforming Education: From Innovation to Implementation, October 10–12, 2011; agenda available at www.purdue.edu/discoverypark/learningcenter/conference-2011/.

student research using NASA's "vomit comet" (a reduced-gravity aircraft that simulates the zero gravity environment of space).[26]

Such collaborations are especially valuable because tribal colleges are generally isolated geographically, which is why they have very good video conferencing capabilities. These collaborations allow students and faculty to become involved with research, which puts the science in context for the students, gets them interested, and connects them with universities. What makes tribal colleges good for modifying curriculum is that, unlike larger universities with engineering schools that have a lot of inertia for changing curriculum, the tribal colleges are usually smaller and can change quickly.

Effective Interventions (2): Constructive Controversy

Karl Smith, professor of cooperative learning in engineering education, Purdue University, began by seconding McKenna's recommendation to think carefully about the alignment of the educational approaches with the goals and assessment of the project. The project team members should also think about what they want students to know and be able to do. His sense was that, among other aims, the project team wanted to foster conversations among a variety of audiences inside and outside academia, as well as more deep and critical thinking about climate change, engineered systems, and society.

Based on those goals the project might make good use of a pedagogical method known as *constructive controversy*, which Smith helped develop in the 1970s and 1980s. It is designed to help students understand an issue and its arguments from all sides through a cooperative effort. Students are assigned a position that they prepare, present, and defend; then they switch sides and drop the advocacy component; finally they either come up with a recommendation or identify the best arguments on all sides. Investigators at the University of California at Los Angeles also researched and tried this approach, which they called *controversy with civility*.

The approach adheres to the following guidelines for "skilled disagreement":

- Define the decision as a mutual problem, not as a win-lose situation.
- Be critical of ideas, not people (confirm others' competence while disagreeing with their positions).
- Separate one's personal worth from others' reactions to one's ideas.
- Differentiate before trying to integrate.
- Pay attention to others' perspectives before refuting their ideas.
- Give everyone a fair hearing.
- Follow the canons of rational argument.

In support of the utility of explicitly engaging with differences, instead of avoiding or automatically refuting them, Smith quoted Alfred Sloan. As chair of General Motors, Sloan once concluded an executive meeting called to consider a major decision by saying, "I take it we are all in complete agreement on the decision here.... Then I propose we postpone further discussion until our next meeting to give ourselves some time to develop disagreements and perhaps gain some understanding of what the decision is all about."

Smith called on the project team to expand its focus beyond outputs to a research and innovation approach to engineering education, in which research evidence leads to changes in both theory and

[26] Information about this NASA feature is available at www.nasa.gov/audience/forstudents/brainbites/nonflash/bb_home_vomitcomet.html.

practice. He showed a diagram neatly illustrating "the innovation cycle of educational practice and research," in which educational practice identifies and motivates questions and ideas, which lead to educational research that results in answers and insights that in turn help improve educational practice.[27] His Collaboratory for Engineering Education Research has produced workshops to encourage such research and approaches, and Smith suggested that the CCEP project follow this model by including research in its activities to develop education.

In addition, the project team should address the challenge of how to change engineering faculty. One approach is to encourage them to think of themselves as *designers* rather than imparters of knowledge, an idea he credited to Jim Duderstadt at the University of Michigan. With respect to the project's focus on producing pedagogical resources, he reiterated McKenna's point that simply putting a resource on the web in hopes that people will use it does not work.

Effective Interventions (3): Inquiry-Based Projects

Suresh Dhaniyala, associate professor of mechanical and aeronautical engineering at Clarkson University, described his development of a general engineering class at that university with funding from a NASA grant on Global Climate Change Education (the class was categorized under general engineering science rather than a specific engineering department to reduce administrative burdens associated with department regulations and paperwork). The class was offered in 2010 and 2011, but a lack of institutional support meant there was no provision for teaching assistance nor was the professor's teaching load reduced to compensate for the addition. And because climate science is a quickly changing field, it requires more effort for engineering faculty to stay up to date on the topic. He combined his research with the class topic and thus mitigated the problem and benefited as a teacher.

In designing the course, Dhaniyala and his coteacher Sue Powers, Spence Professor of Sustainable Environmental Systems in the Department of Civil and Environmental Engineering, wanted to make the class appealing to engineering students. To them this meant inquiry-based projects that were student-defined, guided by the professors, and discussed in class. They made the course quantitative to be more relevant to the students.

Students received NASA data on temperatures, precipitation, and other climate measures and were invited to draw their own conclusions. Then the teachers taught the climate science. (This approach—not starting with the climate science until students determined for themselves whether the climate was changing—was suggested by climate literacy experts they consulted.) Dhaniyala and Powers also engaged the students using the controversy with civility approach described by Karl Smith. The second part of the course called for the students to think about how to address climate change, framing it as a problem for which they could develop solutions. The goal was for students to learn that climate change is a subject that can be addressed through engineering and that there are career opportunities to do so.

Dhaniyala presented assessment data based on questions to the students before and after the course. The assessment measured knowledge, behavior, affect, and self-efficacy, and the results showed an increase in all measures after the class. Dhaniyala shared three questions and their results. In response to the first question, "What is the most important problem facing the United States today?," 22 percent of the students identified climate change before the class versus 74 percent after. When asked "Is global warming caused mostly by human activities?," 14.8 percent agreed before versus 82.6 percent after. In the students' answers to the question, "Does climate change only impact future generations?," 18.5 percent said yes before the class, compared to only 4.3 percent after.

[27] The diagram was from Leah H. Jamieson and Jack R. Lohmann. 2009. *Creating a Culture for Scholarly and Systematic Engineering Educational Innovation*. Washington: American Society for Engineering Education.

The experience revealed the following opportunities for similar efforts to teach climate science and policy to engineering students: team teaching, integration of research activities with teaching, and teaching an interdisciplinary set of students and bringing them together to solve problems.

Dhaniyala concluded with some lessons learned:

- Political/administrative considerations at universities need to be addressed.
- External funding was crucial in getting past organizational roadblocks—without it his work would have been much more difficult.
- Materials and syllabi for faculty would be helpful, and his project has plans to produce them.
- Climate science must be incorporated in the engineering curriculum, and would fit best as part of the fundamental engineering background taught to students so that they are aware of the issue when they learn about mitigation, adaptation, and engineered systems.

In the discussion following his presentation, Dhaniyala said that climate change education for engineers would be possible in a general earth sciences/climate change class rather than an engineering class, although an earth sciences course on climate change would provide more climate science background than engineering students might need (in his course only a few weeks of earth science knowledge had been necessary). Asked why female enrollment in the class was so high compared with the much lower percentage of women in engineering at the college (40 percent in the class, 18 percent in the college), Dhaniyala posited that it was a function of the higher number of women in environmental engineering and engineering management and that the class had preferentially attracted people from those areas.

Panel Discussion and Questions

Project team members Liz Cox of Red Rocks Community College, Jon Leydens and Junko Munakata Marr of the Colorado School of Mines, and Ed Berger of the University of Virginia joined the panelists for a discussion.

Aligning Pedagogy to Professional Goals

Berger began by asking the panelists their thoughts about aligning pedagogy with the goals of the profession, not just those of the classroom, observing that climate change education might serve the larger professional goals of increasing diversity and bringing in historically underrepresented students. McKenna thought that tying the project goals with the profession's goals would improve the success of the project.

Smith commented that such alignment, which he referred to as "backward design" (i.e., starting with outcomes rather than pedagogy), might be applicable at the program or even university level now that regional accreditation is shifting to emphasize outcomes, but he warned about pushing pedagogical ideas and strategies too far. McKenna proposed that, instead of "pushing" results out, the project team figure out what the "pull" is for the profession and faculty to change.

Pedagogical Tensions

A question from the audience cited the tension between systems thinking as a pedagogical approach to climate change education and the more common approach in engineering education that is discipline-based and focused on problem solving without systems thinking. Several panelists commented that many faculty members might consider themselves to be systems thinkers, but their system is much smaller than the system for climate change. Delborne suggested that the project might need to push engineering education to teach systems-level thinking through quantitative and qualitative ideas and methods. Berger cautioned that changing how disciplinary topics are taught could raise concerns about accreditation, which would dissuade faculty and programs from adopting the new pedagogies.

Leydens, shifting the conversation, asked the panelists to think about framing and mentioned the risk of linguistic landmines when using the term "climate change." At his university faculty use "energy efficiency," "energy conservation," "energy use reduction," and "sustainability" more often than "climate change." Cox agreed—in her experience with business organizations working on sustainability practice, the terms "global warming" and "climate change" were emotional triggers. Delborne countered that if the team's goal is to teach climate change to engineers, the term and its meaning must be explicitly introduced into engineers' thinking. McKenna proposed that the team frame the topic as a challenge in need of a solution (instead of focusing on the scientific side of climate change), because engineering students tend to be problem solvers.

An audience member said the project should incorporate sociotechnical thinking in engineering education. In response McKenna noted that ABET Criterion H requires that engineering students be taught global, social, environmental, and economic context. Dhaniyala said his class had involved sociotechnical systems thinking by bringing in speakers to talk about policy issues, financial aspects, and social implications. Delborne asked Madsen how sociotechnical systems thinking fit into education at the tribal colleges, and whether they were better positioned to incorporate it. Madsen responded that tribal colleges want their students to think broadly and have knowledge beyond the field they want to pursue as a career, and that such a goal was compatible with sociotechnical systems thinking. Cox pointed out that community colleges are more limited in the changes they can make to curriculum because they have to ensure their classes can be accepted as transfer credit.

Research on Pedagogical Intervention

The final topic for discussion was the incorporation of research in the team's approach to pedagogical intervention. Madsen observed that incorporating research in the pedagogy is very important for tribal colleges because it allows the faculty to be more involved and gives them the time to better develop the courses. Berger added that involving faculty in research opens new intellectual opportunities that are recognized and consistent with the reward process at their institutions. McKenna suggested that it might be better to frame research as scholarship, as the expectation of doing research in engineering education might be too much for engineering faculty.

Online Resources

At a lunchtime discussion during the second workshop, representatives from a number of online sites that feature materials relevant to the CCEP project described their contents:

- **Online Ethics Center for Engineering and Research** – onlineethics.org
  An electronic repository of resources on science, engineering, and research ethics, for engineers, scientists, scholars, educators, students, and interested citizens.

- **Ethics CORE** – nationalethicscenter.org
  An electronic library of resources on ethics in science and the responsible conduct of research; materials available for kindergarten through postgraduate study.

- **Cleanet** – cleanet.org
  A collection of educational resources on climate change and energy topics; materials available for 6th grade through college.

- **Climate CoLab** – climatecolab.org
  A collaborative website that harnesses the collective intelligence of thousands of people from all around the world who submit and comment on proposals on how to address climate change.

The discussion indicated that the sites addressed the needs of different audiences, but that increased interaction would improve their ability to address those needs.

In Summary

Presentations and discussion acknowledged a number of challenges to integrating CC&ES education into education at various levels and into engineering curriculum. However a number of educational interventions that incorporated issues of sustainability, justice, public trust and engagement, and governance into the curriculum with CC&ES were identified and recommended. Some of the suggested interventions were used in formal education and the results were reported so that lessons could be learned for expanding the formal education efforts of the project.

Chapter 6
INFORMAL EDUCATION ON CLIMATE, ENGINEERED SYSTEMS, AND SOCIETY

At the second and capstone workshops speakers described the role of science centers[28] in informal education about climate change, applicable models for informal climate change educational activities, and the use of art to engage and inform the public on the project topics: climate change, severe weather, and their impacts on infrastructure and society.

Science Center Capabilities for Informal Education and Public Engagement

Speakers articulated the specific capacities of science and technology centers in effectively communicating multifaceted information, and described a number of programs to engage the general public and school-aged audiences on climate change, engineered systems, and society as a component of the CCEP collaboration.[29]

<u>Overview</u>

There are more than 350 science centers in the United States, and in 2010 they had more visitors than all professional sporting events and amusement parks combined (Figure 6.1). These centers connect people of all ages with science, technology, engineering, and math (STEM) by providing first-hand—and often hands-on—experience, encouraging curiosity, supporting formal learning, and inspiring and providing opportunities for students at the local level. They can develop content quickly and present diverse views, both expert and nonexpert, and they are viewed by the public as a trusted source of information.

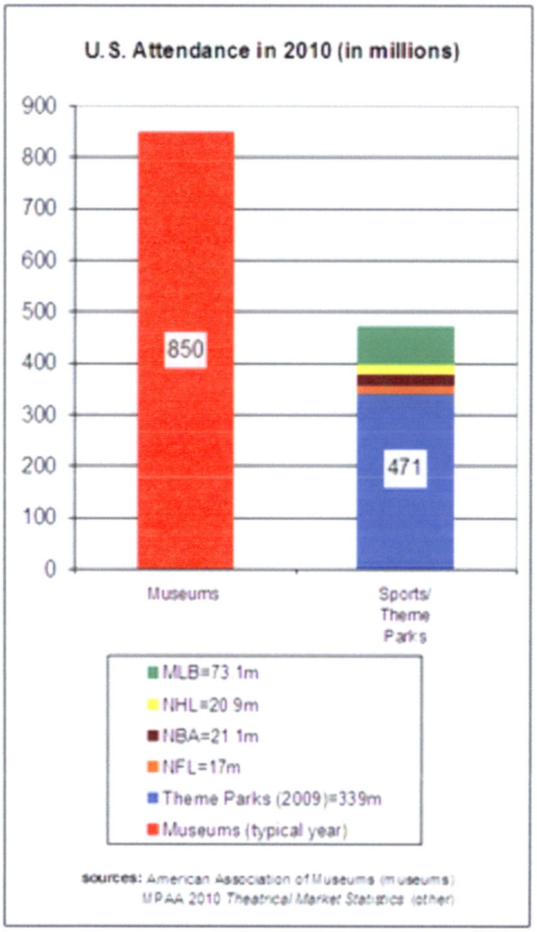

Figure 6-1

Paul Fontaine, vice president of programs at MOS, reviewed the variety of educational resources provided by science centers. Beyond exhibits, museums feature digital media, theater and art activities, curriculum development, and professional development for teachers. They offer forums, lectures, and panels, bringing in experts to meet and talk with visitors; interactive exhibits with three-dimensional objects; demonstrations of science phenomena; and interpretations based on thoughtful discussion and deliberation. Some

[28] In this chapter the terms science center and museum are used interchangeably.

[29] The agenda and slides of these presentations at the second workshop are available at www.nae.edu/Projects/CEES/57196/35146/62343/52752.aspx.

exhibits can be designed for travel to other institutions to extend their impact. The various options take different amounts of time and have different benefits; for example, exhibits typically take three years to create, whereas forums can be put together much more quickly, but reach fewer people.

Fontaine explained that the MOS learning strategy was for people who visit the museum to realize that they inhabit two worlds simultaneously—the natural world and the engineering world—and that the two worlds affect each other. In that light, social justice issues associated with climate change are an exciting new area for science centers to explore.

Informal Nano Education

Rae Ostman, director of national collaborations at Sciencenter in Ithaca, New York, agreed that science centers are very effective at developing students' interest in science and getting them to identify with the scientific enterprise, and she referenced the National Research Council's report on *Learning Science in Informal Environments* (2009).[30]

The focus of her presentation was the Sciencenter's Nanoscale Informal Science Education Network (NISE Network; www.nisenet.org), which offers some clear parallels for informal learning about climate change, engineered systems, and society. In addition to exploring the relationship between science, engineering, and technology and daily life, society, and the environment, NISE efforts include explicit consideration of societal and ethical implications, acknowledging that nanotechnologies have costs, risks, and benefits that cannot always be predicted. The network seeks to increase the capacity of informal education to bring content and experiences to the public on the topic of nanotechnology, and the same could be done for climate change and engineered systems.

NISE Net is a national community of researchers and informal science educators dedicated to fostering public awareness, engagement, and understanding of nanoscale science, engineering, and technology. Its goals are to

- Create a network community to increase capacity in the field:
  - Support partners in engaging the public in nanoscale science, engineering, and technology
  - Form partnerships among informal science education institutions and research centers
- Engage the public through educational experiences:
  - Develop and distribute educational products
  - Raise public awareness and understanding of nano.

NISE activities incorporate the following "strands of learning":

- Develop interest in science through motivation, curiosity, and enthusiasm.
- Promote understanding of science knowledge and the natural world.
- Engage the public in scientific reasoning (e.g., asking and answering questions, evaluating evidence).
- Invite the public to reflect on science—to understand it as a way of knowing.
- Engage the public in scientific practice to achieve greater understanding.
- Enable the public to identify with the scientific enterprise by establishing a level of comfort, knowledge, and interest.

[30] National Research Council. 2009. *Learning Science in Informal Environments: People, Places, and Pursuits*. Washington: National Academies Press.

The network has 14 core partners involved in producing materials, about 100 "second-tier" partners that integrate the education into their regular activities, and about 300 "third-tier" partners that are just starting to incorporate nanotechnology in their curriculum. The core partners are organized both regionally and by educational product, such as exhibits, programs, NanoDays, and other public engagement such as media. NanoDays, an annual event that occurs on the same day at hundreds of science centers across the country, gets nanotechnology information out to the public through kits designed for science centers to create temporary setups; the kits include hands-on activities, public programs, media and graphic materials, marketing materials, and training materials.

Because the topic was so new, the team had to both create products and inspire a desire for the knowledge—factors similar to those that characterize the topic of climate change, engineered systems, and society. The NISE team created products aimed at different audiences, with different levels of detail, such that the widest audience had shorter, less intense experiences and smaller audiences had longer, more in-depth experiences. The team used a three-part iterative, collaborative process: (1) expert input to find interesting ideas and ensure that they are accurately presented; (2) peer review to ensure that products achieve educational goals, are well crafted, and represent best practices; and (3) visitor evaluation with target audiences to ensure that experiences are accessible, engaging, and educationally effective.

NISE products that emphasize the societal implications of nanotechnology and that may be of interest to the CCEP are five posters and three activities for inclusion in the NanoDays kits; a webpage that lists and responds to staff and visitor questions about societal implications (with Arizona State University); and a short video aimed at young adults that is a parody of 1950s education films. To encourage dissemination of the materials the team created an online catalogue for informal science education professionals and a public website that details their products and events.[31]

Communicating Climate Change

Kate Crawford, project manager for the NSF-funded Communicating Climate Change (C3) project developed and led by the Association of Science and Technology Centers (ASTC), described the project and lessons learned that would be relevant to informal education about climate, engineered systems, and society.[32] The ASTC is a professional association of over 400 science centers and museums throughout the world. It became involved in climate change communication after a Yale project demonstrated that even Americans who accept that climate change is real and caused by humans do not see it as a problem for their generation or for them personally.[33] Educators proposed science center programming around local indicators of climate change to help address this communication problem.

The C3 project consists of partnerships with research institutions and science centers to develop cases and programming around local indicators. To be able to modify the programming as the project progressed, the partners developed citizen science programs, community conversations, and public dialogues. The citizen science programs engage laypeople in the collection and analysis of scientific data to advance research.

The project team found that when picking a local indicator it was crucial that the item be easily linked to climate change and be emotionally important to the people in the community. It was also beneficial, but often difficult, to make sure the indicator lent itself to an existing citizen science project to ensure the data

[31] More information about the NISE network and its products is available at www.nisenet.org/.

[32] Information about the project is available at www.birds.cornell.edu/citscitoolkit/climatechange/projects/c3.

[33] Information about the Yale Project on Climate Change Communication and its surveys is available at http://environment.yale.edu/climate-communication/.

would be used. One science center, the Maryland Science Center, opted not to follow the local indicator approach and instead created products on urban heat island effect; Crawford mentioned the project because of its relevance to climate change and to underscore the difficulty of communicating the subject.

Partnerships with researchers at academic institutions are most successful when the researchers get something out of participating, such as use of the data in their own research, and when the public understands how their efforts are helping. The science centers get the benefit of having an expert readily available on a topic that is important to the museum and to community education programs. But building these partnerships is difficult because researchers who volunteer often are not aware of the time commitment involved. Several successful partnerships have been with graduate students (who valued the free data collection).

Criteria for Model Programs and Products

David Sittenfeld, manager of the Forum Program at the Museum of Science in Boston, proposed a model of a partnership in informal science education for the CCEP project based on the C3 and NISE Network programs. Such a model should

- leverage the expertise that the CCEP network has to offer;
- be diverse in its content, in terms of type of educational activities used, engineering questions explored (e.g., from various engineering disciplines), and climate change topics examined (including differences among climate zones); and
- reach diverse audiences and communities, both geographically and institutionally, and consider cultural, political, and social dimensions.

Regionally based programs would be anchored by a local informal science education institution, with a mix of expertise from the CCEP network to develop the content. Resulting materials should connect the process of doing science with the engineering design process.[34] The curriculum should also communicate the inherent uncertainty in decision making around emerging science and technology as well as short-term versus long-term considerations for the engineering, social, and ethical challenges, including their different impacts on various communities and stakeholders and on different timescales. Last, the educational materials should be (1) available online, (2) open-source so that other educators can figure out how to create their own similar resources, and (3) adaptable so that others can customize them for different audiences and uses.

Programs and Activities at Museums and Science Centers

Leading science museums in the United States have developed innovative educational activities on issues related to climate change and infrastructure. Participants at the capstone workshop heard presentations from the Boston Museum of Science (MOS), the Science Museum of Minnesota (Saint Paul), the Chabot Space and Science Center (Oakland, California), and the Koshland Science Museum of the National Academy of Sciences (Washington, DC).[35]

In his introduction of the session, David Rabkin, director for current science and technology at MOS, defined informal education in general as a means to help develop and nurture the skills and habits of informed and critical thinking, and described science museum education as a forum for "free-choice"

[34] As an example of a project that does this effectively, he cited NASA's project, Beginning Engineering Science and Technology Curriculum (www.nasa.gov/audience/foreducators/best/).
[35] The agenda and video and slides of speakers' presentations are available at https://www.regonline.com/builder/site/tab2.aspx?EventID=1155563.

learning, as compared to compulsory K–12 education. Because museums rely on a voluntary audience, they have to appeal to the public, and to that end they make use of interactive exhibits, art and visual representation, and extramural activities. In addition, museums are proactive, oriented toward timely and forward-thinking topics, new methods for education, and new ways of partnering with their communities and other organizations. He concluded by reporting that most people view science centers as safe, trustworthy, and welcoming to diverse groups of people, and this makes them good places for discussions of climate change.

Boston Museum of Science

David Sittenfeld described several MOS activities conducted as part of the CCEP project. A series of presentations, titled "Behind the Headline: Engineering for Our Changing Climate," looked at decision making about engineered systems in a changing climate, examining an issue in the local news headlines and translating it for a public audience. Issues were drawn from three case studies—on increases in sea level, extreme precipitation events, and extreme temperatures—in the 2011 publication of the *Massachusetts Climate Change Adaptation Report*.[36] The activity engaged museum guests in considering climate change in public policy decisions about infrastructure and making decisions under conditions of uncertainty about future environmental changes.

The first case asked people to decide how they would have planned for the siting of the Deer Island Wastewater Treatment Plan on an island at sea level, based on current knowledge of the impacts of climate change. The second case involved designs and plans for building the Spalding Rehabilitation Hospital on a piece of land vulnerable to floods and storms. The third looked at potential effects of extreme temperatures on the Massachusetts transit system—on passenger comfort and disrupted operations if tracks buckle, energy demands reduce energy availability, and storm water infiltrates the system. Sittenfeld reported that participants were very interested in the discussions and were in fact more willing to think about the impact of climate change on infrastructure than they were about climate change more generally.

Evaluations revealed that participants were not interested in learning about the evidence of climate change or the reasons for it but rather wanted to discuss and make decisions about how to handle the impacts of climate change on infrastructure. This successful activity was conducted for only three weeks at the museum, but it was converted and made available online through the museum's biweekly podcast series on science and technology.[37]

In a second initiative, MOS researched existing science museum efforts and activities on climate change. Of 69 such programs or exhibits at science museums across the country, only 7 addressed the impact of climate change on engineered systems and infrastructure. Most were about evidence for climate change or ecosystem impacts. Half of the programs used case studies to convey impacts, but again only 7 presented multiple factors and perspectives, which are important for decision making.

The third MOS program for the project was a 7-week-long workshop for high school–aged youth. The weekly workshop focused on planning for healthier cities in a changing climate and included research on mapping urban heat islands and assessing air quality in public spaces. Students crafted research questions in collaboration with expert mentors from public health, urban planning, and community research, reviewed background information and case studies relevant to their questions, and then gathered and

---

[36] The report is available at www.mass.gov/eea/docs/eea/energy/cca/eea-climate-adaptation-report.pdf.
[37] The podcasts are available on the MOS website (http://legacy.mos.org/events_activities/podcasts/podcasts_archive&d=5392).

analyzed data from the Boston area. They also developed discussion points on the issues for a forum involving peers, policymakers, community members, and scientific stakeholders.

Katie Behrmann, MOS program fellow, presented the plans for a fourth CCEP project activity, a public forum on sea level rise in the Boston Harbor. Even conservative predictions of sea level rise in the Boston area would compromise significant landmarks such as the MIT campus, the Museum of Science, the Aquarium, and Fenway Park. Superstorm Sandy demonstrated the urgency of this concern, making climate change and infrastructure an ideal topic for such a forum, which focused on adaptation to rather than evidence for climate change. The forum was designed to get people thinking about who or what is impacted, how long a policy or structure should last, how much risk people want to plan for, and who pays for the adaptation. Using case studies on the Deer Island Sewage Treatment Plant and Spaulding Rehabilitation Hospital, participants looked at three aspects of the city likely to be impacted by sea level rise: infrastructure, historical sites, and vulnerable neighborhoods. As they explored ways to plan for and adapt to changes in sea level rise, participants were asked to consider the perspectives of commuters, long-time business owners, public housing residents, city planners, engineers, and students. They then presented their plans and discussed them with the other forum participants.

Science Museum of Minnesota

Patrick Hamilton, program director at the Science Museum of Minnesota, endorsed the NISE Network program (described above) as an ideal model for a similar large-scale education program on climate change and infrastructure, and then presented his understanding of the goals for the CCEP effort on climate change and infrastructure:

- Raise awareness among citizens, policymakers, and decision makers of the implications of a changing climate;
- Increase the willingness and capacity of citizens, decision makers, and policymakers to support climate change resiliency; and
- Pursue climate change resiliency strategies that have other societal benefits.

Science museums can support these three goals by *framing*, *convening*, and *catalyzing* conversations. Exhibits, workshops, forums, and conferences are all examples of framing devices: their planning determines what is included in (and excluded) from consideration of a topic. Convening can be done passively (e.g., through the marketing of new exhibits, museum visitors' choice of which exhibit to view) and actively (e.g., through forums whose time, place, and audience are planned). Catalyzing spurs action on a topic. When it comes to any kind of "wicked [or messy] problem," these three things—framing, convening, and catalyzing—are critical to the outcome of efforts to successfully address the problem.

Hamilton described the museum's Future Earth Strategic Initiative and its activities exploring water, energy, food security, climate, and, most recently, climate change resiliency. Among these was a scenario-planning workshop that framed, convened, and catalyzed a discussion on climate change and resiliency to inform the city of Saint Paul's sustainability office. The event convened 22 people from the city, state, and federal government, academia, and private nonprofits to develop four plausible scenarios for the implications of economic variables, demographic changes, and climate change between 2012 and 2040. The activity catalyzed the cities of Saint Paul and Minneapolis to include climate change adaptation in their plans for the future, whereas previously their focus had been only on mitigation efforts.

Chabot Space and Science Center

Eric Havel, education manager at the Chabot Space and Science Center, reviewed its programs and activities on climate and adaptation: (1) community conversations (e.g., forums); (2) citizen science projects, in which citizens help collect data on the climate and its possible impacts; and (3) climate literacy programs, such as a teacher training program on climate change that also provides curricular materials for teachers to use in the classroom.

Chabot activities demonstrate that climate change hits close to home, with maps showing impacts on California's weather, agriculture, snowpack, and plant growth. One project explores whether sword fern frond lengths are changing over time and, if so, whether this indicates a change in moisture levels (rainfall and/or fog drip) due to climate change. Citizen scientists are enlisted to measure fern frond numbers and lengths of local populations and compare them with others in the redwood ecosystem.

Marian Koshland Science Museum of the National Academy of Sciences

Jeanne Braha Troy, program officer at the Koshland Science Museum, explained that the museum's target audience is science-interested adults and its mission is to help people use science to solve problems, drawing from intellectual materials produced by National Academies expert committees. The museum transforms the intellectual knowledge into engaging experiences through exhibits or programs that help visitors develop critical thinking and problem-solving skills as well as a sense of self-efficacy in their decision making on science and engineering topics.

The museum's current exhibit on climate change, Earth Lab, asks visitors, "Climate change is happening but what can one do about it?" Based on a suite of Academies reports on *America's Climate Choices* and *America's Energy Future*, the exhibit uses data visualization to allow visitors to drill down into the science, see how much energy is being used in different countries, and then use a mitigation simulator to see how different policy choices will contribute to carbon reduction goals. Evaluations have shown that this activity helps people realize that there are a number of options and that there isn't just one solution or silver bullet to reducing carbon emissions. People are encouraged to think about what they can do on a practical level to help reduce their carbon footprint.

The museum also does extramural and collaborative programs to extend its outreach. For example, a CCEP project called the Climate and Urban Systems Partnership (CUSP; www.cuspproject.org) aims to make climate change more relevant to individuals by moving from the global to the neighborhood and community level. The partnership plans to engage people to think about climate change, mitigation, and adaption as they go through their day at various places in the urban environment and infrastructure—in other words, to engage them with what matters to them in their daily life. People's experiences are then connected to urban systems and climate change so they can see how climate change will directly affect them and understand the impacts of their decisions.

Engaging the Public through Art

A session at the capstone workshop examined visual and non-classroom-based methods for engaging the public about long-term, local, and often invisible changes in the environment, in some cases specifically due to climate change.

Jody Roberts, director of the Center for Contemporary History and Culture at the Chemical Heritage Foundation (CHF), spoke about a new project, Sensing Change, designed to communicate and visualize local environmental change through the work of artists, featuring images of what crisis might look like when it happens. The exhibit was imbedded in both the museum and community, through public installations and programming such as the CHF Distillations podcast, science cafés, and public conversations between scientists and artists. A selection of works illustrated the array of media and creative approaches to art in the service of science.

Figure 6-2

A public installation by Andrea Polli, titled Particle Falls, is a projection on a building of a 60-foot light visualization of real-time air quality data, giving viewers immediate information on particulate pollution levels based on laser light scattering measurements (Figure 6.2). Another artist, Roderick Coover, created panoramic animated videos of the Philadelphia river estuary and overlaid it with maps, charts, and diagrams showing the predicted effects of rising water on historic and modern sites. Artist Stacy Levy created the Calendar of Rain, in which glass bottles collected rainwater during a 24-hour period and were then displayed on a shelf to present a physical "bar graph" of rainfall over several weeks (Figure 6.3). And landscape artist Diane Burko uses USGS data to show changes in landscapes she has been painting for many years.

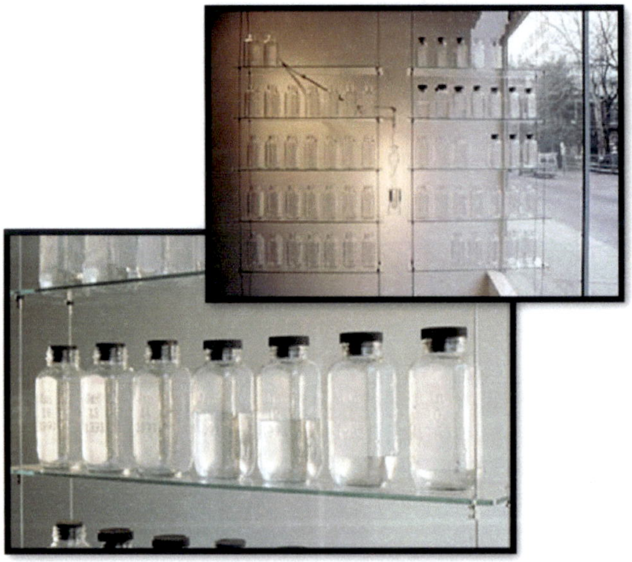

Figure 6-3

Roberts concluded by showing a video by artist and interventionist Eve Mosher on her project called the HighWaterLine, a project to help New Yorkers visualize the impact of climate change on their city.[38] In 2007—five years before superstorm Sandy—she chalked 70 miles of Manhattan and Brooklyn that would be vulnerable to mega floods if climate change continued (Figure 6.4). The sight of Mosher drawing the chalk line drew people to her in conversations about climate change and its impacts. The project also revealed a number of infrastructure and public utility facilities that are located below or at the chalked line (which corresponds to the 100-year flood mark) and will be unusable when flooding reaches it. Mosher characterized the project as an opportunity for public leaders, community groups, experts, and people living in the affected communities to work together to be more resilient and responsive. For

Figure 6-4

[38] The video is available at http://vimeo.com/58422367.

complicated issues that may seem too large to grasp, she said, art can create simplicity and personalize the events, and reach people in a way that is more humanized than science, technology, and politics.

Kira Appelhans, a landscape architect, described a project commissioned by New York City's Museum of Modern Art (MOMA) to revision flood zone infrastructure around the New York City Harbor. Rising Currents[39] was primarily a landscape architectural design project, with a focus on increasing soft infrastructure or living coastline features such as dunes, salt marshes, and oyster beds. The section of coastline involved in the project included a petroleum refinery, shipping docks, and residential living. Proposed measures were (1) installation of a land berm to cap and contain contamination from the petroleum refinery, (2) transformation of petroleum storage tanks by cleaning them (using algae) and converting them for biofuel production, and (3) creation of large glass "jacks" from recycled glass to slow storm surges and reduce the size of waves in the harbor (Figure 6.5). The design plans were shared at public events that were well attended and the plans were then converted into an exhibit at MOMA. Appelhans reported that the project engaged people who would not ordinarily have been interested in climate science information by making the impacts realistic and personalized. Furthermore, the exhibit prompted the city of New York to incorporate sea level change in its flood maps in 2011. City officials also organized and met with 22 communities in coastal areas to discuss sea level rise and storms and their impacts.

Figure 6-5

Stacy Levy, a sculptor from Spring Mills, Pennsylvania, made the case for the inclusion of the arts in interdisciplinary teams with scientists and engineers and showed examples of such collaborations to reclaim built areas in ways both functional and attractive for better land and water management. Explaining that visual metaphors can be very effective for explaining how the natural world works, she described projects she has created to communicate science through art and to incorporate weather and natural processes in cities. In addition to the Calendar of Rain, Levy has done a number of projects that reveal the actions of water through rain, rivers, and tides. One uses large colorful flower petals fastened around coastal piers to visually display the tide level (Figure 6.6). Another uses a curtain of plastic buoys hanging from strings to demonstrate the level and speed of a river (Figure 6.7).

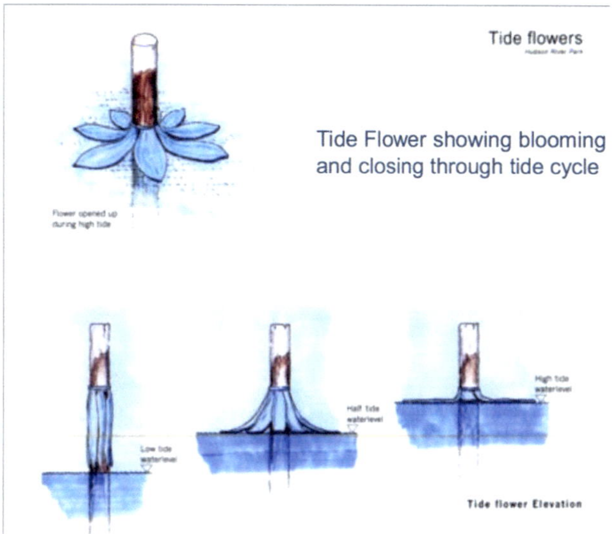

Figure 6-6

[39] Information about Rising Currents is available at www.moma.org/explore/inside_out/category/rising-currents#description.

Levy called on engineers and scientists to recognize the utility of a well-crafted artistic metaphor to make information and solutions more visible and understandable to the public.

Figure 6-7

In Summary

Presentations and discussions described a wide range of interventions and methods for conducting informal CC&ES education. The interventions were designed to engage people in complex ethical, policy, and engineering decisions while making them approachable topics to the general public. The interventions highlight the issues the public faces regarding climate change and engineered systems, while also discussing and encouraging consideration of the cross-cutting themes.

Chapter 7
PERSPECTIVES OF ENGINEERING PROFESSIONAL SOCIETIES, BUSINESS AND INDUSTRY, LOCAL GOVERNMENT AND NATIVE AMERICANS

Input from educators, both formal and informal, was crucial to the Climate Change Education Partnership, but the project partners also realized the importance of broadening its outreach to entities that support, hire, rely on, or prepare engineers in other ways. The project therefore engaged a variety of communities in the discussions of the intersection between education and action on climate change and engineered systems. This chapter presents the perspectives, suggestions, and efforts of engineering professional societies, business and industry, local governments, and Native Americans.

Engineering Professional Societies

William Kelly, director of external affairs at the American Society of Engineering Education, moderated a session on the activities and educational priorities of engineering professionals on climate change.[40]

ABET Accreditation

William Wepfer, chair of the Engineering Accreditation Committee, vice president for education at the American Society of Mechanical Engineers, and chair of the Woodruff School of Mechanical Engineering at Georgia Institute of Technology, spoke about ABET, a federation of 30 professional engineering and technical societies that works to accredit engineering educational programs and to promote quality and innovation in education.

ABET accredits educational programs based on outcomes that correspond with eight general criteria: (1) students, (2) program educational objectives, (3) student outcomes, (4) continuous improvement, (5) curriculum, (6) faculty, (7) facilities, and (8) institutional support. The most important of these are the educational objectives and student outcomes. The student outcomes criteria are 11 abilities that all engineering students are expected to have mastered by the time of graduation (Box 7.1).

Box 7.1
ABET Student Outcomes

a) An ability to apply knowledge of mathematics, science, and engineering
b) An ability to design and conduct experiments, as well as to analyze and interpret data
c) An ability to design a system, component, or process to meet desired needs within realistic constraints such as economic, environmental, social, political, ethical, health and safety, manufacturing, and sustainability
d) An ability to function on multidisciplinary teams
e) An ability to identify, formulate, and solve engineering problems
f) An understanding of professional and ethical responsibility
g) An ability to communicate effectively
h) The broad education necessary to understand the impact of engineering solutions in a global, economic, environmental, and societal context
i) A recognition of the need for, and an ability to engage in, lifelong learning
j) A knowledge of contemporary issues
k) An ability to use the techniques, skills, and modern engineering tools necessary for engineering practice

[40] The agenda and slides of these presentations at the second workshop are available at www.nae.edu/Projects/CEES/57196/35146/62343/52752.aspx.

In thinking about how to incorporate climate change in engineering curriculum, it might be better to pick an educational outcome that is either relevant across engineering programs or, alternatively, targeted to specific disciplines or a new discipline. Wepfer also pointed to the flexibility of the education outcomes and curriculum criteria, specifically the requirement for a culminating major design that incorporates appropriate engineering standards and multiple realistic constraints—as might characterize a project to address climate change.

American Society of Civil Engineers (ASCE)

Sustainability has been a long-term interest of the American Society of Civil Engineers (ASCE) and was incorporated into its code of ethics in 1996. The group defines sustainability as "a set of environmental, economic, and social conditions in which all of society has the capacity and opportunity to maintain and improve its quality of life indefinitely without degrading the quantity, quality, or availability of natural, economic, and social resources." Climate change is considered a component of ASCE's sustainability efforts (although some members do not believe that it is occurring).

Richard Wright, chair of the ASCE Sustainable Infrastructure Education Subcommittee, described the Sustainable Infrastructure Education (SIE) Program, whose objectives are to provide both the body of knowledge for certification of sustainable infrastructure professionals and the knowledge basis for Env**is**ion™ (www.sustainableinfrastructure.org), a sustainable infrastructure project rating system developed through a partnership of ASCE, the American Council of Engineering Companies, and the American Public Works Association.

The SIE modules (available online or in development) cover

- fundamentals of sustainable engineering;
- sustainable project management;
- community participation ;
- land use and ecological issues;
- water, air, light, mobility, noise, and waste; and
- assessment of project life cycle impacts.

Specific courses in the Fundamentals of Sustainable Engineering program (www.asce.org/fsecourse) are designed to address the following:

- Introduction: fundamentals of sustainable engineering and professional certification in sustainable engineering
- Transformational projects: examples of and rationale for the transformational approach to sustainable engineering
- Trends and issues: economic, environmental and social concerns for sustainability.
- Earth systems: the Earth's natural life support systems and how engineers apply principles of sustainability to preserve them
- Five capitals: natural, produced, human, social, and financial
- Social factors: the community
- Social factors: individual behavior
- Sustainability quadrant: human development and its ecological footprint
- Moving toward sustainability: addressing sustainability in infrastructure sectors
- Project pathway and performance: doing the right thing and the thing right
- Life cycle cost/benefit assessment
- Life cycle environmental assessment

- Environmental policies, regulations and innovation
- World view for sustainable development
- Delivering sustainable projects
- Leadership perspectives

Envision is similar to the US Green Building Council's LEED program for sustainable buildings but also encompasses sustainable infrastructure, which includes communications, energy, transportation, and water and wastewater treatment facilities. The system has 10 categories for rating sustainable infrastructure projects:

- Project pathway contribution
- Project strategy and management
- Communities—long- and short-term effects
- Land use and restoration
- Landscapes
- Ecology and biodiversity
- Water resources and environment
- Energy and carbon
- Resource and waste management
- Access and mobility

("Project pathway contribution" includes consideration of both the people to be served by the infrastructure and those who will be affected by it, such as neighboring communities.)

The Technologies for Carbon Management Project (http://fscarbonmanagement.org) was initiated in 2008 by five engineering societies (AIChE, AIME, ASME, ASCE, and IEEE) with support from the United Engineering Foundation to address the roles of the engineering community in mitigation of and adaptation to climate change. It provides information about the characterization and measurement of greenhouse gas emissions, potential effectiveness (e.g., time and cost) of options, and knowledge gaps and other barriers to implementation. The project has held two meetings for interested professionals.

In addition to the ASCE initiatives, Wright cited the NSF-sponsored Center for Sustainable Engineering: 250 engineering faculty members have attended its summer institutes for training to incorporate sustainability in the engineering curriculum.

Wright concluded with two observations. First, that there is nothing unusual about engineers dealing with climate because they have been doing so ever since they have been building structures that resist the forces of nature. Second, the lack of connection between climate models and weather models is a problem for engineers because they need quantitative data on the impacts of more severe weather on engineered structures. In the past, engineers used historical weather data to understand weather forces and probability, but climate scientists have proven that these weather trends are variable and that historical weather data will not be sufficient for engineering designs in the future.

Engineers Canada/Ingénieurs Canada

Engineers Canada/Ingénieurs Canada is a national body of 12 provincial and territorial associations that regulate the practice of engineering in Canada. It accredits undergraduate engineering programs (similar to ABET in the United States) and facilitates common approaches for professional qualifications, professional practice, and ethical conduct. The organization decided early on not to get into the why's of climate change, but to accept that the climate is changing and that engineers need to deal with it. A National Climate Change Action Plan was established in 2003 that included communication and outreach, education, continuing professional development, adjustments to engineering practices, networking of scientists and engineers, advice to government, and funding arrangements.

Engineers have a responsibility to assess situations and manage risk; to develop and/or revise policies, standards, and guidelines; and to perform due diligence in addressing changing climate for engineering work. A multijurisdictional and multidisciplinary effort is needed to address the challenges of climate change, especially among climate scientists, engineers, geologists, and operations and maintenance specialists. Engineering-related consequences of climate change include premature deterioration, higher maintenance and operation costs, and reduced performance and life span/life cycle.

David Lapp, manager, professional practice, Engineers Canada, Secretariat, Public Infrastructure Engineering Vulnerability Committee (PIEVC), described a collaborative project with the Canadian Standards Association on climate change and infrastructure. The objective was to examine solutions, assess the current state of curriculum, determine practitioners' awareness through a survey, and produce findings and recommendations. The project explored four infrastructure categories that affect many of the major engineering disciplines: buildings, energy, transportation, and water.

A survey of practicing engineers to establish a baseline for awareness of issues related to climate change showed that 82 percent of them believed a changing climate will affect their engineering decisions in the near future—and 73 percent felt they needed much more information to enable them to incorporate its impacts into their engineering practice. The survey also asked about their familiarity with tools and techniques related to climate change: although 77 percent were "at least somewhat familiar" with "encouraging energy efficiency and low emission solutions," only a third said the same about "designing infrastructure that can be modified over time with the impacts of climate change" (33 percent) and "identifying locations that may be vulnerable to climate change impacts and then modifying designs accordingly" (35 percent).[41] Thus although most infrastructure engineers accept that climate change will affect their practice in the future, few actively factor this into their decisions. The survey results have significant implications for infrastructure since its design life can and often does exceed 50 years.

The project participants assumed that an existing but underused body of knowledge relevant to climate change could be integrated in the curriculum, but in fact they found very little specific climate change content and no engineering courses dedicated to the topic. They determined that risks, codes, standards, and frameworks, as well as discussions of the triple bottom line, are key areas for incorporation in climate change curriculum. And case studies and examples are crucial because they provide opportunities for engineers to get down to specifics.

The final report recommended six topic areas that should be included in undergraduate curriculum: (1) climate change science; (2) decision-making processes around economic, environmental, and societal issues; (3) climate change impacts and adaptations; (4) risk; (5) public policy and regulatory frameworks (codes and standards); and (6) psychology relating to decisions, perceptions, and behavior. The report also emphasized that flexibility and a modular approach are essential for successfully integrating these components in engineering curriculum. For practicing engineers, the report recommended that they be encouraged to customize their learning experience, that climate change topics be promoted in continuing education programs, that alternative educational methods be offered on climate change, and that engineers receive training in the areas of risk, decision making, uncertainty, and emergency planning.

Engineers Canada collaborated with Natural Resources Canada on the Public Infrastructure Engineering Vulnerability Committee (PIEVC), which was tasked with (1) overseeing a national engineering assessment of the vulnerability of public infrastructure to climate change in Canada, (2) facilitating the development of best engineering practices for adaptation to climate change impacts, and (3) making recommendations for reviews of infrastructure codes and standards. The committee developed a 5-step risk assessment protocol developed for use by qualified engineering professionals to assess the risk of

[41] Further results from this survey are available from Dr. Lapp at Engineers Canada.

climate change in infrastructure. Two phases of 1-day educational workshops were organized to train engineers in using the assessment protocol and educate them about climate change; the phase 1 workshops were targeted to associations of Engineers Canada and phase 2 to governments and municipalities across Canada on a cost-recovery basis.

Engineers Canada has also developed a syllabus for a 36-hour course on climate change. It is designed for continuing education but can also be used for graduate education.

Lapp concluded that there is already an established body of knowledge that can be used by engineers to address climate change, and that it must be incorporated in educational initiatives at universities or through continuing education to get the information out to mainstream engineering practice.

Business and Industry Perspectives

What do engineering businesses consider the underlying principles, skills, and experiences that will prepare future engineers to effectively meet the challenges of climate change in the practice of engineering? Five industry representatives were invited to address this and related questions in a panel at the second workshop. The panelists were Keith Williams, chief technology officer in the Materials, Corrosion, and Environmental Technology Division at Science Applications International Corporation (SAIC); Jonathan T. Malay, director, Civil Space and Environment Programs at Lockheed Martin; John Carberry, an independent consultant and retired director of environmental technology at DuPont; Laurens van der Tak, vice president and water resources engineer at CH2M Hill; and William Flanagan, director of the Ecoassessment Center of Excellence at General Electric (GE) (he was unable to attend but sent comments that the panel's moderator shared). Panel moderator Andres Clarens, assistant professor of environment and water resources engineering at the University of Virginia, facilitated a discussion based on questions sent to the panelists in advance, focusing particularly on (1) corporate responses and actions, (2) characteristics and skills for the engineers hired, and (3) case studies for teaching climate change in engineering.

Corporate Responses and Actions on Climate Change

Van der Tak reported that CH2M Hill became interested and involved in issues of climate change because of its corporate ethic of sustainability, which required a good understanding of climate science to ensure that engineered designs could withstand climate changes. So the company invested in the development of software enabling access to global circulation models and the latest climate change projections, and trained some of its engineers to use the software and interpret the information. In its climate change adaptation work, CH2M Hill has focused on infrastructure master planning, design, and construction. Whether clients request it or not, the company considers the long-range climate picture in designs for multimillion-dollar infrastructure that may be sensitive to climate changes. CH2M Hill does this because it is the right thing to do and, as Van der Tak put it, engineering for climate change adaptation is not any different than doing good engineering work.

At Dupont, Carberry's responsibility was to figure out what environmental factors, including climate change, would affect the chemical industry, specifically its chemical plants and customers. The impacts he cited as needing consideration were energy consumption at corporate buildings and the productivity of customers' agricultural products and practices.

Characteristics and Skills of Hired Engineers

Malay pointed out that while Lockheed Martin is known for its work on airplanes and military technologies, but is also very involved in engineering for environmental science and environmental monitoring. However, they are an engineering firm and thus when it comes to hiring engineers the company cares primarily about their technical engineering skills rather than if they know about climate science.

Van der Tak explained that CH2M Hill has divided its business related to climate change into two parts: adaptation and mitigation. For hiring in adaptation the company looks for civil and environmental engineers as well as agricultural engineers and environmental scientists. Most of those hired have a master's degree, and some a PhD; bachelor's-level engineers are rarely hired. The company also provides training for its engineers in risk and risk assessment methods.

Carberry made the case that the industry needs engineers that start out as high school students who understand that climate change is real and will have wide-ranging impacts on society. When they become trained engineers they will be better prepared to understand how climate change will impact the engineering work—for instance, that there will be wider swings in temperature and therefore a need for products that help mitigate the temperatures, such as air conditioners and insulation. Engineers must also be able to think critically and take a long-term view of their work.

Clarens reported that GE seeks graduates with an ability to think on a systems basis. They should pay attention to global trends and drivers, emerging regulations, industry standards, customer requests and requirements, green public procurement policies, competitive behavior, and product ingredients.

Carberry, Van der Tak, and Williams agreed that universities and educators should incorporate climate change in sustainability education and that more engineers should be educated about sustainability. To that end, Carberry called for faculty access to resources that will enable them to identify how climate change will affect the teaching of their subject area. He also observed that recent graduates going into industry should be aware that their concerns about a company's environmental responsibility will be better received if they present them in terms of impacts on the business and its relations with customers.

All the panelists agreed that continuing education is important to ensure that working engineers have current information about climate change impacts. They can access such information through professional societies, online, or in university classes.

Cases Studies for Effectively Exploring Climate Change in Engineering

Representatives from GE suggested cases on (1) corn-based ethanol; (2) the cost of ownership and comparative environmental burdens for a hybrid versus diesel versus gas vehicle; (3) cost analysis of electricity generation for nuclear, gas turbine, coal, and renewables; and (4) business and root cause analysis for energy engineering disasters (e.g., oil leaks and spills).

Carberry proposed two cases: a comparison of the Kyoto and Montreal protocols and why one failed and the other succeeded; and a review of the US sulfur oxides and nitrogen oxides cap and trade system.

According to Van der Tak, cases on the climate impacts of just about any type of public infrastructure project would be useful. They would need to be very site specific, but would reveal broadly applicable concerns such as how long the facility was supposed to last when thinking about sea level rise, increasing storm surges, and changing power demands.

Malay suggested a case on the California freshwater distribution system and seconded the idea of comparing vehicle choices.

Local Government Perspectives

At the County Level

Katy Simon, manager of Washoe County, Nevada, spoke about her local government's engagement in sustainability efforts. The county commission's strategic objective is the sustainability of the community's financial, social, and natural resources. Although some elected officials do not believe that climate change is occurring, support for sustainability has crossed political parties there. It was a focus on financial sustainability that brought the elected officials to the table to discuss sustainability in all aspects of the city—water conservation, flood control, snow pack monitoring, water reuse, environmental building practices, wildfire risk management, habitat and air quality management, and local food production.

Simon cited the importance of public participation, accurate and transparent communication, and trust in decision makers. She also said that citizens will support expensive infrastructure funding and policy changes if they know the local government is doing everything it responsibly can before resorting to expensive changes. Last, a characteristic specific to Washoe County that has helped in its sustainability efforts is the residents' close connection to the natural environment.

Nancy Gassman, natural resources administrator for Broward County, Florida, described climate-related challenges in South Florida and actions at multiple levels of governance. The county is experiencing weather events such as extreme rainfall, drought, cold, and high temperatures, some of them at the same time; in 2011, for example, significant rainfall on Halloween coincided with high tide and overwhelmed storm drains. Coastal flooding and storm surges are threatening drinking water because of salt water intrusion as well as power plants vulnerable to flooding from only a 1-foot rise in sea level.

To address and plan for these events Broward County (1) has done an inventory of the county-owned infrastructure at risk and assessed climate change impacts, (2) is encouraging green and climate impact resistance construction practices, and (3) is adopting adaptation standards that require consideration of climate change and sea level rise in the design of all new public buildings. The county works at every level of government to address climate issues, to plan to maximize the useful life of existing infrastructure in the short term, and to incorporate climate considerations into longer-term master planning and land use decisions.

At the City Level

Jonathan Koehn, environmental affairs manager for the City of Boulder, Colorado, described challenges and actions Boulder is dealing with, including wildfires, floods, and energy plans. He agreed with Simon's suggestions and comments, and added that local government officials share strategies for making changes and moving toward a more sustainable community. When taking action it is important for local officials to connect with the community about what it values and to communicate that changes in the environment are the new norm.

He cited the following areas in which impacts from climate change will be felt in Boulder and other US cities:

- Water and energy demand/costs
- Vulnerable populations (cooling, working conditions, inadequate housing, etc.)
- Built environment (design considerations and extreme weather impacts)
- Livability/health (air and water quality, recreation, habitat)

- Economy (tourism, transportation, business disruption)

The city learned firsthand about reaching the local community with information about climate change impacts when wildfires encroached. City officials connected what had been abstract and global perceptions of climate change with the local experience by pointing out that, while climate change did not cause the wildfires, it made the community more vulnerable and susceptible to them.

Kevin Burke, city manager for Flagstaff, Arizona, explained the development of the city's Resiliency and Preparedness Initiative. The initial question in the planning was "How can the city of Flagstaff reduce its vulnerability and build local resilience to climate variability and climate-related disasters?" But because of resistance to climate change in their community and government, the topic was reframed as emergency planning, which made people more receptive to the project. The effort considered leadership, management, and operations perspectives, and sought ways to avoid the unimaginable; manage the unavoidable; capture future natural hazard scenarios; continue the government's mission to protect life, health, property, and infrastructure; and reduce the severity of risk. The project also identified primary systems that would see impacts from climate change and then mapped those to the key government planning areas (Box 7.2).

Box 7.2

Primary System	Key Planning Areas
Emergency Services	Police and Fire Services, EMS, Disaster Response, Public Works
Energy	Energy Delivery and Assurance, Energy Demand and Cost
Forest Health	Forest Management, Wildlife and Vegetation, Public Infrastructure
Public Health	Public Health Infrastructure, People, Public Services
Stormwater	Buildings, Infrastructure
Transportation	Public Transportation, Transportation Infrastructure, Public Access, Rail, Airport
Water	Water Treatment Quality, Water Resources, Water Infrastructure

One way to effectively communicate the impact of predicted temperature changes was to identify a nearby community that was already more arid and had the environmental characteristics that resembled those projected for Flagstaff in the future. Environmental conditions in the nearby community were noticeable and serious enough for community members to realize the impact of even a small change in temperature.

Burke concluded with some lessons learned for similar efforts: (1) follow a team approach with broad representation; (2) look first at things you can control; (3) adapt the process as you go along; (4) focus on implementation; (5) identify opportunities to prepare; and (6) concentrate on impacts to avoid getting bogged down in questions of whether, how, and why the climate is changing.

Sam Lipson, director of environmental health for the City of Cambridge Public Health Department in Massachusetts, discussed public health impacts associated with climate change:

- Heat-induced power loss and patient surges from heat-related illness
- Gastrointestinal and respiratory illness from pathogens, mold, bacteria, and asthma
- New or expanded vector-borne risks of West Nile virus, Lyme disease, and Dengue fever
- Emotional and psychological effects such as stress, depression, loss of community, and grief
- Flood-induced loss of water and power at medical facilities and limited access to medical services
- Heightened risks for home-bound residents
- Postflooding consequences such as sewage-tainted water, mold, and bacteria

The challenges of addressing these public health concerns involve planning for rare but high-impact events and engaging agencies and utilities in work across sectors and disciplines. Cambridge has begun taking action by conducting a vulnerability assessment of at-risk populations, at-risk areas, essential services, public infrastructure, and transit and power systems. It is also working to ensure public health and public safety preparedness and to assess and anticipate health burdens now and in the 50-year time frame.

Native American Perspectives

Tracey LeBeau, director of the US Department of Energy's Office of Indian Energy Policy and Programs (OIEPP), described the office's efforts to address the priorities of Indian tribes for energy development in the context of climate change. The office was authorized in 2005, received a budget in 2009, and develops its priorities based on feedback from over 250 tribes. Public meetings with tribes indicate strong interest in renewable and clean energy, so the office is establishing programs in finance and markets as well as ways to develop infrastructure for sustainable economies, areas in which tribes need training.

Tribal communities face challenges from subpar infrastructure, but a 2012 report about tribal lands indicated substantial potential for large-scale commercial renewable projects, and in 2013 the office and tribes began to focus on small-scale development.[42] Conversations are just beginning. With massive tribal land areas, what are the potentials for carbon capture and storage? What technical assistance is needed for very small Alaskan communities to develop a resilient infrastructure that can be completely off-grid? The tribes need to examine the relationship of climate change and their energy choices, but that examination has barely begun.

A panel then provided perspective from three Native American organizations; Patricia Mariella, director of the American Indian Policy Institute, ASU, chaired the session. Ann Marie Chischilly, executive director of the Institute for Tribal Environmental Professionals (ITEP), reported that climate change has a disproportionate impact on 566 tribes and Alaskan natives for reasons that range from relative poverty to impacts on subsistence living and sacred sites, and cited a number of challenges facing particular tribes and regions. ITEP provides training, assistance, and educational resources to tribes on climate change issues, focusing on adaptation planning, tribal climate change profiles, traditional ecological knowledge, participation in the national climate assessment, and the First Stewards organization. Its environmental education outreach program reaches students from kindergarten through college. It partners with numerous organizations and tribes to host the Tribal Clean Energy Resource Center, which provides energy planning and technical and policy analysis to the tribes, and sponsors internships and professional development.

Pilar Thomas, deputy director of the DOE Tribal Energy Office, noted that traditional energy resources and transmission are largely owned by nontribal entities. New energy sources can change this ratio and enhance the capacity for energy security and climate change mitigation and adaptation in Indian country. She listed a number of capabilities that are essential to develop this potential, from planning to emergency response and funding. Technological and organizational capabilities are also necessary, particularly if opportunities for adaptation are to be recognized and incorporated into infrastructures.

Jose Aguto, legislative secretary for sustainable energy and environment at the Friends Committee on National Legislation, reminded the audience that tribal governments are beginning to exercise their rights and need to be consulted as governments. However, tribal authority within their jurisdictional boundaries

[42] December 2012. DOE Office of Indian Energy *Developing Clean Energy Projects on Tribal Lands Data and Resources for Tribes* (www.nrel.gov/docs/fy13osti/57048.pdf).

is very difficult to exercise, as is evident in questions of jurisdiction concerning nonmembers' land within tribal boundaries, and special requirements from federal agencies such as the Bureau of Indian Affairs.

He cautioned against overuse of generalizations about tribes but acknowledged common issues and priorities. Two characteristics fundamental to indigenous people grow out of their place-based values: their close relationship with their natural resources and the desire to transmit these resources to future generations. To uphold their values many tribes exercise best practices in natural resource management; for instance, traditional ecosystem practices underlie salmon harvest and renewable forestry programs. The tribal world view emphasizes the links between the protection of ecosystems and prosperity, happiness, and survival, and Aguto called on the larger community of experts and interested citizens to develop cooperative efforts to support and expand these opportunities.

The discussion after the panel covered numerous topics—provision of engineering education in and for tribal colleges; development of STEM capacity for Native Americans in nontribal colleges and universities; incorporation of Native American perspectives in the focus on climate change, engineered systems, and society; and development of new models that incorporate education about systems management and policy in a wide range of undergraduate and graduate programs to provide students from all disciplines a systemic approach to management decisions about scientific and technical training.

In Summary

Broadening participation in this CCEP Phase I project educated project participants about the approaches each took to the topic. They were able to articulate their organizational needs and priorities, results from relevant research and related activities, and the challenges and opportunities that the issues posed for them. Placing these discussions in the context of sociotechnical systems demonstrated that solutions would demand the engagement of all of these communities in responding to the multi-faceted interactions among climate change, engineered systems, and society. All project participants are grappling with these issues, and with the associated questions about sustainability, governance, justice, and public trust that any interventions pose. The summaries presented in this chapter and in the report allow readers to examine the results of this process – what has been accomplished and more important perhaps what next steps might look like. Professional societies, business and industry, local governments, and Native American tribes are developing responses that others might consider.

Appendix A

Workshop Agendas

Workshop on Climate, Society, and Technology

Huntington Room at the Beckman Center of the National Academies

June 7-8, 2011

Agenda

Background: The goal of the CCEP Phase I project on "Climate Change, Engineered Systems, and Society" is to develop conceptual and educational frameworks and networks of change agents to promote effective formal and informal education for engineering students, policymakers and the public at large. These activities should address, visibly and systematically, issues of climate and engineered systems, including issues of governance, sustainability, justice, and public engagement and trust. The goal of the workshops component of the project is to lay the foundations for the project partners—the National Academy of Engineering, Arizona State University, Boston Museum of Science, Colorado School of Mines, and University of Virginia-Charlottesville—to use in developing these frameworks and networks.

In the intersection of climate, engineered systems, and society, it is the second term in this triumvirate that provides important and under-recognized challenges and opportunities for our examination. The implications of how engineered systems interact with climate for engineers and the public must be emphasized in the project. The planning effort for this project includes two workshops: one focusing on the interactions between climate and socio-technological systems, and a second one on the educational dimensions of this interaction between climate and those systems.

The first workshop focuses primarily on issues of adaptation and mitigation for climate and engineered systems, where these systems are understood as complex socio-technical systems with significant political, cultural, economic and ethical dimensions. It also pays attention to larger scale climate interventions such as geo-engineering. The second workshop focuses on the implications for engineering and public education of incorporating the interactions between climate change and engineered systems.

The first workshop is scheduled for June 7-8, 2011; the second workshop for October 18-19, 2011. For both events, the day prior to the workshop (June 6 and October 17 respectively) consists of a project planning meeting with project members; days two and three are the public workshops.

Available members of the project team and external advisory board may meet also for a short post- workshop review just following the event.

Tuesday, June 7, 2011

7:30 – 8:25 **Breakfast**

8:25 – 8:35 **Call to Order: Rachelle Hollander, NAE CEES**

8:35 – 10:30 **Session I: Interactions-Defining the Problems**

In this opening session, speakers will present views about climate and its interaction with engineered systems understood as socio-technical systems, from the varied perspectives of their expertise and experience. The session will review the contributions that those perspectives make to identifying and understanding the problems facing engineered systems in society. Much research and many reports identify problems expected from the likely range of interactions among climate, engineered systems and societies, and some recommend solutions. More than a few consider problems of sustainability as an environmental rather than a social issue. Relatively few consider or critically explore associated issues of governance, sustainability in social contexts, justice, and public engagement and trust. In their talks, speakers are invited to explore the ways in which scientific, engineering, political and social interventions and priorities can, do, and should influence the interactions of climate, engineered systems, and society, and how these influences are likely to affect the success of programs and recommendations.

Moderator: **Juan Lucena**
Liberal Arts and International Studies; Colorado School of Mines

Speakers: **James McCarthy**
Biological Oceanography; Harvard University
Science Perspectives

Jay Golden
Center for Sustainability & Commerce; Duke University
Business and Engineering Perspectives

Ann Bostrom
School of Public Affairs; University of Washington
Public Perspectives

Respondents: **Joseph Herkert**
School of Applied Arts and Sciences; Arizona State University

Jason Delborne
Liberal Arts and International Studies; Colorado School of Mines

10:30 – 10:45 **Break**

10:45 – 1:00 **Session II: Interventions-Examining the Range of Socio-technological Responses**

Adaptation? Mitigation? Geo-engineering? Other Large Scale Interventions? All of the above? Often, discussions about responding to climate change focus on one or more of these options and involve projections about potential costs and benefits. Speakers in this session will probe further on the social justice dimensions of these options, e.g., the kinds, likelihood and distribution of potential benefits, costs, risks, and harms from the range of options under discussion. Also considering issues of governance, sustainability, and public engagement and trust, the panelists should summarize and assess positions that have been taken about these interventions, their potential likelihood, and estimations of those associated consequences and their distribution. They should consider how cultural and societal norms and priorities would be likely to influence results.

Moderator: **Junko Munakata Marr**
Environmental Science and Engineering: Colorado School of Mines

Speakers: **Edward Rubin**
Environmental Engineering and Science; Carnegie Mellon University
Mitigation Strategies – Potentials and Problems

Jackie Kepke
Water Portfolio Management; CH2M Hill
Engineering Perspectives – Towards Structural Change

David Daniel
President's Office; University of Texas at Dallas
Adaptation of Technological Systems

Alan Robock
Department of Environmental Sciences; Rutgers University
Geoengineering Potentials and Myths

Respondents: **Kathryn Johnson**
Division of Engineering; Colorado School of Mines

David Slutzky
School of Engineering and Applied Sciences; University of Virginia

1:00 – 2:00 **Lunch**

2:00 – 3:00 **Session III: Panel on Cross-Cutting Themes**

Moderator: **Deborah Johnson**
Science, Technology, and Society; University of Virginia

Panelists: **Joe DesJardin**
President's Cabinet; Saint John's University
Justice

Paul Thompson
Agricultural, Food and Community Ethics; Michigan State University
Sustainability

Susanne Moser
Institute for Marine Sciences; University of California-Santa Cruz
Governance, Trust, Public Engagement

3pm – 4:30pm: Group Breakouts

This session consists of four small group breakouts that will address each of these topics in relationship to the presentations and discussions in prior sessions, and report back to a roundtable/plenary about what we know, and what we need to know, based on the results.

A. Governance (Emerald Bay Room)
B. Justice (Laguna Room)
C. Sustainability (Huntington Room)
D. Public Trust and Engagement (Irvine Cove Room)

Group A - Governance
Facilitator: **David Sittenfeld**
Forum Program; Museum of Science, Boston
Rapporteur: **Borna Kazerooni**
Engineering and Applied Science; University of Virginia

Group B - Justice
Facilitator: **Joseph Herkert**
School of Applied Arts and Sciences; Arizona State University
Rapporteur: **Jon Leydens**
Liberal Arts and International Studies; Colorado School of Mines

Group C - Sustainability
Facilitator: **Helene Hilger**
Civil and Environmental Engineering; UNC-Charlotte
Rapporteur: **Jen Schneider**
Liberal Arts and International Studies; Colorado School of Mines

Group D - Trust, Public Engagement
Facilitator: **Paul Fontaine**
Education; Museum of Science, Boston
Rapporteur: **Liz Cox**
Institute for Sustainability in Education; Red Rocks Community College

4:30 – 5:30 **Reports From Breakouts**
Facilitator: **Rachelle Hollander**
NAE CEES

5:30 – 6:00 **Closing Session**

Wednesday, June 8, 2011

8:00 – 9:00 **Breakfast**

9:00 – 11:00 **Session IV: Education**

This plenary will brainstorm ideas about the implications for education that have come from the prior sessions and informal interactions among workshop participants. The goal of this session is to help us map stakeholders and issues to be considered in workshop II in October which will focus exclusively on education. These considerations should address where limited investments are likely to provide the greatest payoff for a Phase II implementation project.

Chair: **David Rabkin**

Current Science and Technology; Museum of Science, Boston

11:00 – Noon **Session V: What We've Learned**

In this session, the co-principal investigators of the Phase I CCEP award will highlight the initial take-home messages from the workshop and ask the participants for their comments, suggestions, criticisms, and additional thoughts.

Chair: **Clark Miller**

Consortium for Science, Policy & Outcomes; Arizona State University

Noon **Workshop Adjourns**

Networking Educational Priorities for Climate, Engineered Systems, and Society

House of Sweden, Washington DC
October 18-19, 2011

AGENDA

Project Focus and Goals: The goal of the Climate Change Educational Partnership Phase I project on "Climate Change, Engineered Systems, and Society" is to develop a conceptual and educational framework and a network of change agents to promote effective formal and informal education for engineering students, policymakers and the public at large. The project should address, visibly and systematically, issues of climate and engineered systems, including governance, sustainability, justice, and public engagement and trust. The goal of the workshops component of the project is to lay the foundations for the project partners—the National Academy of Engineering, Arizona State University, Boston Museum of Science, Colorado School of Mines, and University of Virginia-Charlottesville—to use in developing the implementation plan for the second phase.

The project assumes that the role of engineered systems vis-à-vis climate and society provides important challenges and opportunities for formal and informal engineering education in classrooms, public forums, and science museums and centers, and those educational programs need to address both technical and societal issues. The implications of the interactions of engineered systems with climate—for engineers, engineering, and the public, must be recognized.

NAE Project Workshops: Considerable research and many reports identify problems expected from interactions among climate, engineered systems and societies; and some recommend solutions. More than a few consider problems of sustainability, as an environmental rather than a social issue. Relatively few consider or examine associated issues of governance, sustainability in social contexts, justice, and public engagement and trust. This project invites participants to explore the ways in which the separation of technical from social issues may affect the success of formal and informal educational programs and recommendations, and how to overcome the divide so as to increase the likelihood of success.

The first project workshop in June 2011 focused on the interactions among climate and social and technological systems. The upcoming workshop on October 18-19, 2011, at the House of Sweden in Washington, DC will focus on education about these interactions. The day prior to the workshop (October 17) consists of a project planning meeting with project team and external advisory board (EAB) members; days two and three are the public workshop. Available members of the team and advisory board may meet also for a short post-workshop review just following the event.

Day One: Tuesday, October 18, 2011

8:30-9:15am Session I: Welcome and Introduction to the Program
This session provides a project overview and status report on the Phase I activities to date, with 5-minute slide presentations from the team leaders.
9:15-11:15am Session II: Effective Interventions in Undergraduate Engineering Education
The goal of session II is to educate project participants about engineering education innovations that can improve the process of integrating climate change and engineered systems (CC&ES) in engineering curricula and scale up across multiple institutions.
The session is divided in two one-hour parts. In Part I speakers address specific questions. The speakers are:

- Jason Delborne, Colorado School of Mines (Moderator)
- Ann McKenna, Arizona State University
- Bob Madsen, Chief Dull Knife College
- Karl Smith, Purdue University/University of Minnesota
- Suresh Dhaniyala, Clarkson University

Part II is a panel-format dialogue to explore the answers further; three or four project representatives will join the speakers for a dialogue about these questions and answers. Audience members will submit questions to a moderator who will present them to the group for responses.

- Edward Berger, University of Virginia- Charlottesville
- Liz Cox, Red Rocks Community College
- Jen Janacek Hartman, United Tribes Technical College
- Jon Leydens and Junko Munakata Marr, Colorado School of Mines

11:15am-noon: Session III: Engineering in the K-12 Curriculum, A Review Richard Duschl, Penn State University

Noon-1pm Lunch

1-1:45pm Plenary Welcome

Introduction: John Ahearne, NAE, Chair, CEES Advisory Group Speaker: Charles Vest, President, National Academy of Engineering

1:45-3:45pm Session IV: Informal Education, Science Center Capabilities and Public Engagement

This session will explore the role that science and technology centers play in the educational community, their institutional strengths and limitations in communicating multifaceted information, and present a model for engaging the general public and school-aged audiences in the topic of climate change, engineered systems and society intended to function within and through the context of the larger CCEP collaboration. It consists of two parts: a panel overview followed by an open space exercise to explore the merits of key aspects of science center engagement.

- Paul Fontaine, Vice President of Programs, Museum of Science, Boston (Moderator)
- Kate Crawford, Project Manager, Communicating Climate Change, Association of Science and Technology Centers, Washington, DC
- Rae Ostman, Director of National Collaborative Projects, Sciencenter, Ithaca, NY
- David Sittenfeld, Program Manager, Forum Program, Museum of Science, Boston

3:45-4pm Break to go to Breakout Groups

4-5:30pm Breakouts

First Day Breakout Groups will consider the following (reporting back to a plenary):

- Ways to enhance undergraduate engineering curricula
- Community and tribal college programs
- K-12 education
- Informal education and public engagement
- Public policy education
- Outreach, dissemination, special projects

(Some breakout groups may consider several topics in the course of discussion. Organizers reserve right to rearrange these sessions based on expressions of interest and program changes).

Adjourn for Day

Day Two: Wednesday, October 19, 2011

8-8:30am Continental Breakfast

8:30-9:15am Report back from Breakouts

9:15-11:00am Session IV: Institutional and Professional Society Initiatives

In this panel session, speakers will provide information about their activities regarding educational priorities for climate, society, and technology. The general discussion will encourage audience members to identify the work other organizations have been doing that addresses these issues and associated opportunities for networking.

- William Kelly, American Society for Engineering Education (Moderator)
- William Wepfer, ABET
- Helene Hilger, University of North Carolina Charlotte
- Dick Wright, American Society of Civil Engineers, Founder Societies' Carbon Management Project
- David Lapp, Engineers Canada/Ingénieurs Canada

11-11:10am Break

11:10-1pm: Corporate Perspectives on Engineering and Education on Climate, Engineered Systems, and Society

The premise here is that engineers should be trained to prepare for addressing issues of climate change. Businesses that employ engineers are well equipped to provide insights into their thinking about these issues in the context of the demands they face and the technological and organizational challenges they see ahead. The panel will focus on what employers of engineers perceive as the underlying principles, skills, and experiences that will prepare future engineers to effectively meet the challenges of climate change in the practice of engineering. The session will consist of two parts. Each of the participants will make a brief introduction, followed by a facilitated discussion.

- Kristina Hill, PhD Associate Professor of Landscape Architecture at University of Virginia (Facilitator)
- Keith Williams, Chief Technology Officer, Navy Research and Engineering Division, SAIC
- Jonathan T. Malay, Director Civil Space & Environment Programs, Lockheed Martin
- John Carberry P.E., Independent Consultant (DuPont retired)
- William Flanagan, PhD, GE Global Research
- Laurens van der Tak, P.E., D.WRE, Water Resources & Ecosystem Management, CH2M HILL

1-2 pm Lunch and Roundtable on Outreach and Dissemination – A brainstorming session on how to work with different electronic dissemination outlets to promote project goals.

Panelists include: Josh Bishoff and Megan O. Hayes, Ethics CORE; Representative from Engineering Pathways (tentative); Frank Niepold, NOAA and Tamara Ledley, TERC – Cleanet.org; Simil Raghavan, Onlineethics.org. Questions to address:

What are the goals of these sites? How might they connect with one another, what audiences will each reach, what additional audiences might we need to reach, etc?

1-3pm: Breakouts

Second day breakouts will consider the following (reporting back to a plenary):

- What academic administrators, deans, and center directors can do
- The role of professional societies
- The politics of climate change
- Corporate engagement
- Outreach, dissemination, special projects

(Some breakout groups may consider several topics in the course of discussion. Organizers reserve right to rearrange these sessions based on expressions of interest and program changes.)

2-3:45pm Reports from Breakouts

3:45-4pm: Break

4pm Closing Session – Next Steps

5pm Adjourn

Climate Change and America's Infrastructure: Engineering, Social and Policy Challenges

January 28-30, 2013

Agenda

Sunday January 27th

4:30-6-00pm Registration Open (Lobby)

Monday January 28th

7:00am Registration Opens - Breakfast Buffet

INTRODUCTION AND OVERVIEW: CLIMATE CHANGE, CLIMATE ADAPTATION, AND INFRASTRUCTURE VULNERABILITY

8:00am **Welcome**

- **Clark Miller,** Associate Director, Consortium for Science, Policy & Outcomes, Arizona State University
- **Rachelle Hollander**, Director, Center for Engineering, Ethics & Society, National Academy of Engineering

Keynote Presentations

8:45am **Kathy Jacobs**, Assistant Director for Climate Adaptation and Assessment, White House Office of Science and Technology Policy

9:45am **Daniel R. Cayan**, Researcher, Climate Atmospheric Science and Physical Oceanography (CASPO), Scripps Institution of Oceanography

10:45am *break*

11:00am **Thomas Wilbanks**, Corporate Research Fellow, Climate Change Science Institute, Oak Ridge National Laboratories

Lunch Speaker

Noon - 1:00pm **Gerald Galloway**, Glen L. Martin Institute Professor of Engineering, University of Maryland

ASSESSING THE PROBLEM

Keynote Presentations: Engineering Perspectives

1:00pm **Kathy Freas**, Global Water Resources Director, CH2MHill

1:30pm **David Lapp**, P.Eng., Manager, Professional Practice, Engineers Canada, Secretariat, Public Infrastructure Engineering Vulnerability Committee (PIEVC)

2:00pm Discussion

2:30pm *Break*

3:00pm **Panel 1: Policy and Governance Challenges and Strategies**

- **Elisabeth Graffy,** Professor of Practice and Senior Sustainability Scientist, Arizona State University (chair)
- **Kristin Baja**, Hazard Mitigation Planner at City of Baltimore, Office of Sustainability
- **Thomas Birkland**, Professor, School of Public and International Affairs, North Carolina State University
- **Patricia Mariella**, Director, American Indian Policy Institute, Arizona State University
- **Jennie C. Stephens**, Associate Professor of Environmental Science and Policy, Department of International Development, Community and Environment, Clark University

4:30pm **Panel 2: Engineering, Justice, and Human Rights**

- **Rachelle Hollander**, Director, Center for Engineering, Ethics & Society, National Academy of Engineering (chair)

- **Barbara Rose Johnston**, Senior Research Fellow, Center for Political Ecology, University of California - Santa Cruz
- **Byron Newberry**, Professor Mechanical Engineering, Baylor University
- **Donna Riley**, Associate Professor of Engineering, Smith College

6:00pm Adjourn for Day/Registration Closes

(dinner on your own—looking for folks to dine with? Meeting in the lobby at 6:30)

Tuesday, January 29th

7:00am Registration Opens - Breakfast Buffet

CASE STUDIES IN INFRASTRUCTURE VULNERABILITY AND ENGAGING POLICY AND THE PUBLIC

Keynote Presentations: Sea Level Rise and Storm Surge

8:00am **Greg Kiker**, Associate Professor, Agricultural and Biological Engineering, University of Florida

8:30am **Robert Lempert**, Director, Frederick S. Pardee Center for Longer Range Global Policy and the Future Human Condition, RAND Corporation

9:00am Discussion

9:30am *Break*

10:00am **Panel 3: Local Government Solutions**

- **James Svara**, Professor, School of Public Affairs, and Director, Center for Urban Innovation, Arizona State University (chair)
- **Kevin Burke**, City Manager, City of Flagstaff, AZ
- **Nancy Gassman,** Environmental Protection and Growth Management Department, Broward County, FL
- **Jonathan Koehn**, Regional Sustainability Coordinator, City of Boulder, CO
- **Sam Lipson,** Director of Environmental Health, Public Health Department, City of Cambridge, MA
- **Katy Simon**, County Manager, Washoe County, NV

Noon **Panel 4: Engaging the Public in Climate Vulnerability and Adaptation** *(over lunch)*

- **Jody Roberts**, Director, Center for Contemporary History and Culture, Chemical Heritage Foundation (chair)
- **Kira Appelhans**, Rising Currents / Working Waterline, NYC
- **Stacy Levy**, Sculptor, Spring Mills, Pennsylvania
- **Eve Mosher**, artist and interventionist, NYC (video presentation)

1:30pm **Panel 5: Colorado River Water Resources**

- **Armin Munevar**, CH2MHill (chair)
- **Kay Brothers**, former Deputy General Manager, Southern Nevada Water Authority
- **Chuck Cullom**, Geologist/Hydrologist, Central Arizona Project
- **Carly Jerla**, Co-Study Manager, Colorado River Basin Water Supply and Demand Study, U.S. Bureau of Reclamation
- **Clifford Neal**, Water Resources Advisor, City of Phoenix, AZ

3:30-4:00pm *Break*

Native Perspectives Keynote

4:00pm **Tracey LeBeau**, Director, U.S. Department of Energy's Office of Indian Energy Policy and Programs

4:30pm **Panel 6: Native American Perspectives**

- **Patricia Mariella**, Director, American Indian Policy Institute, Arizona State University (chair)
- **Jose Aguto**, Friends Committee on National Legislation

- **Ann Marie Chischilly**, Director, Institute for Tribal Environmental Professionals
- **Pilar Thomas,** Deputy Director of DOE Tribal Energy Office
- Representative from the Gila River Indian Community

6:00pm **Adjourn for Day/Registration Closes**

(dinner on your own—looking for folks to dine with? Meeting in the lobby at 6:30)

Wednesday, January 3oth

7:00am Registration Opens - Breakfast Buffet

EDUCATIONAL CHALLENGES

8:00am **Panel 7: Informal Science Education**

- **David Rabkin**, Director, Current Science and Technology, Museum of Science, Boston (co- chair)
- **David Sittenfeld**, Program Manager, Forum, Museum of Science, Boston (co-chair)
- **Katie Behrmann**, Programs Fellow, Museum of Science, Boston
- **Jeanne Braha Troy**, Program Officer, Marian Koshland Science Museum of the National Academy of Sciences
- **Patrick Hamilton**, Program Director, Environmental Sciences and Earth-System Science, Science Museum of Minnesota
- **Eric Havel**, Education Manager, Chabot Space & Science Center 9:45am *break*

10:00am **Panel 8: Engineering Education**

- **Tom Seager**, Associate Professor, School of Sustainable Engineering and the Built Environment, Arizona State University
- **Mary Ann Curran**, Life Cycle Assessment & Sustainability Consultant, BAMAC Ltd.
- **Helene Hilger**, Associate Professor Emerita in Civil and Environmental Engineering, UNC Charlotte

11:30am **Concluding Remarks**

- **Rachelle Hollander**, Director, Center for Engineering, Ethics & Society, National Academy of Engineering
- **Clark Miller**, Associate Director, Consortium for Science, Policy & Outcomes, Arizona State University

Noon Conference Adjourns 12:30pm Registration Closes

12:45pm CChESS Core Planning Group meets

5:00pm CChESS Core Planning Group adjourns

This conference is part of the National Science Foundation Climate Change Education Partnership: Climate Change, Engineered Systems and Society project led by the National Academy of Engineering. The Conference is organized by the Consortium for Science, Policy & Outcomes at Arizona State University.

For more information on the Climate Change Education Partnership: Climate Change, Engineered Systems and Society project, see http://www.onlineethics.org/Projects/CCEP.aspx

For more information on the National Academy of Engineering, see http://www.nae.edu/

For more information on the Consortium for Science, Policy & Outcomes at Arizona State University, see http://www.cspo.org.

Speaker biographies are available on the conference website at http://www.regonline.com/builder/site/default.aspx?EventID=1155563

Appendix B

Participants Lists

Climate Change, Engineered Systems, and Society Workshop on Climate, Society, and Technology - Workshop Attendees

June 7-8, 2011

Rebecca Allen
Graduate Student
Arizona State University
rebecca.allen@asu.edu

Ann Bostrom
Associate Dean of Research
University of Washington
abostrom@uw.edu

Grace BullTail
Engineering Instructor
United Tribes Technical College
gbulltail@uttc.edu

Beth Cady
Program Officer
National Academy of Engineering
ecady@nae.edu

Liz Cox
Director, RISE
Red Rocks Community College
Liz.Cox@rrcc.edu

David Daniel
President
University of Texas at Dallas
dedaniel@utdallas.edu

Jason Delborne
Assistant Professor
Colorado School of Mines
delborne@mines.edu

Joe DesJardin
Vice Provost
Saint John's University
jdesjardins@csbsju.edu

Jay Golden
Director, Duke Center for Sustainability & Commerce
Duke University
Jay.Golden@Duke.edu

Joe Herkert
Associate Professor
Arizona State University
joseph.herkert@asu.edu

Helene Hilger
Director, IDEAS
hhilger@uncc.edu

Kristina Hill
Associate Professor
University of Virginia
keh3u@virginia.edu

Rachelle Hollander
Director, Center for Engineering, Ethics, and Society
National Academy of Engineering
rhollander@nae.edu

Deborah Johnson
Professor of Applied Ethics
University of Virginia
dgj7p@virginia.edu

Kathryn Johnson
Assistant Professor
Colorado School of Mines
kjohnson@mines.edu

Nathan Kahl
Sr. Program Associate
National Academy of Engineering
Nkah@nae.edu

Borna Kazerooni
Graduate Student
University of Virginia
bk4z@virginia.edu

Jackie Kepke
Global Technology Leader
Water Portfolio Management
CH2M Hill
jacqueline.kepke@CH2M.com

Dhara Kothavala
Graduate Student
Arizona State University
lkothava@asu.edu

Christine Lan
Graduate Student
University of Virginia
ccl2m@virginia.edu

Jon Leydens
Associate Professor
Colorado School of Mines
jleydens@mines.edu

Juan Lucena
Associate Professor
Colorado School of Mines
jlucena@mines.edu

Junko Munakata Marr
Colorado School of Mines
Associate Professor
junko@mines.edu

Jim McCarthy
Professor of Biological Oceanography
Harvard
jmccarthy@oeb.harvard.edu

Clark Miller
Associate Professor
Arizona State University
clark.miller@asu.edu

Sharlissa Moore
Graduate Student
Arizona State University
sharlissa.moore@asu.edu

Susi Moser
Research Associate
University of California-Santa Cruz Institute for Marine Sciences
promundi@susannemoser.com

Alan Robock
Professor
Department of Environmental Sciences
Rutgers University
robock@envsci.rutgers.edu

Ricky Rood
Professor of Atmospheric Oceanic, and Space Sciences
University of Michigan
rbrood@umich.edu

Bob Roth
CEO, Big Green Zero
bobr@biggreenzero.com

Ed Rubin
Professor of Environmental Engineering and Science
Carnegie Mellon University
rubin@cmu.edu

Jen Schneider
Assistant Professor
Colorado School of Mines
jjschnei@mines.edu

David Sittenfeld
Forum Program Manager
Boston Museum of Science
dsittenfeld@mos.org

David Slutzky
Research Associate Professor
University of Virginia
slutzky@virginia.edu

Walter Staveloz
Director, International Relations
Association of Science-Technology Centers
wstaveloz@astc.org

Bill Tang
Professor, Biomedical Engineering
UC-Irvine
wctang@uci.edu

Paul Thompson
W. K. Kellogg Chair in Agricultural, Food and Community Ethics
Michigan State University
thomp649@msu.edu

Linda Tseng
Graduate Student
UC-Irvine
ytseng@uci.edu

Networking Educational Priorities for Climate, Engineered Systems, and Society – Workshop Attendees October 18-19, 2011

John Ahearne
Chair, CEES Advisory Group, NAE
Sigma Xi
ahearne@sigmaxi.org

Rebecca Allen
Graduate Research Assistant
CNS Biodesign Fellow
Arizona State University
rebecca.allen@asu.edu

Frazier Benya
Program Officer
National Academy of Engineering
fbenya@nae.edu

Edward Berger
Associate Professor of Mechanical and Aerospace Engineering
University of Virginia
berger@virginia.edu

Josh Bishoff
Ethics CORE
University of Illinois
bishoff2@illinois.edu

Beth Cady
Program Officer
National Academy of Engineering
bcady@nae.edu

John Carberry
Independent Consultant
University of Delaware (Dupont retired)
johncarberry01@comcast.net

Vivienne Chin
Senior Administrative Assistant
National Academy of Engineering
vchin@nae.edu

Liz Cox
Director, RISE
Red Rocks Community College
Liz.Cox@rrcc.edu

Kate Crawford
Project Manager, Communicating Climate Change
Association of Science and Technology Centers
kcrawford@astc.org

Jason Delborne
Assistant Professor of Liberal Arts and International Studies
Colorado School of Mines
delborne@mines.edu

Suresh Dhaniyala
Associate Professor of Mechanical and Aeronautical Engineering
Clarkson University
sdhaniya@clarkson.edu

Richard Duschl
Waterbury Chair in Secondary Education
Penn State College of Education
Rad19@psu.edu

William Flanagan
Eco-assessment Leader GE Global Research
flanagan@research.ge.com

Paul Fontaine
Vice President of Programs
Boston Museum of Science
pfontaine@mos.org

Jay Golden
Director, Duke Center for Sustainability & Commerce
Duke University
Jay.Golden@Duke.edu

Megan Hayes
Ethics CORE
University of Illinois
mohayes2@illinois.edu

Joe Herkert
Associate Professor of Ethics and Technology
Arizona State University
joseph.herkert@asu.edu

Helene Hilger
Director, IDEAS
hhilger@uncc.edu

Kristina Hill
Associate Professor of Landscape Architecture
University of Virginia
keh3u@virginia.edu

Rachelle Hollander
Director, Center for Engineering, Ethics, and Society
National Academy of Engineering
rhollander@nae.edu

Jen Janecek-Hartman
Tribal Environmental Science
United Tribes Technical College
jjanecekhartman@uttc.edu

Deborah Johnson
Professor of Applied Ethics
University of Virginia
dgj7p@virginia.edu

Bill Kelly
Director of Public Affairs ASEE
W.Kelly@asee.org

Dhara Kothavala
Graduate Research Assistant
Arizona State University
lkothava@asu.edu

Catherine Kruchten
Paragon TEC, Informal Educator
NASA Goddard Space Flight Center
catherine.kruchten@nasa.gov

David Lapp
Manager, Professional Practice
Engineers Canada
david.lapp@engineerscanada.ca

Jon Leydens
Associate Professor of Liberal Arts and International Studies
Colorado School of Mines
jleydens@mines.edu

Andrew Light
Senior Fellow
Center for American Progress
alight@americanprogress.org

Anders Liljeholm
Science Educator
Oregon Museum of Science and Industry
ALiljeholm@omsi.edu

Stephanie Long
Science Live Theatre
Science Museum of Minnesota
slong@smm.org

Juan Lucena
Associate Professor of Liberal Arts and International Studies
Colorado School of Mines
jlucena@mines.edu

Bob Madsen
Professor
Chief Dull Knife Community College
bmadsen@cdkc.edu

Jonathan T. Malay
Director, Civil Space & Environment Programs
Lockheed Martin
Jon.malay@lmco.com

Junko Munakata Marr
Associate Professor of Environmental Science and Engineering
Colorado School of Mines
junko@mines.edu

Ann McKenna
Associate Professor of Engineering Arizona State University
Ann.McKenna@asu.edu

Clark Miller
Associate Professor of Science Policy and Political Science
Arizona State University
clark.miller@asu.edu

Sharlissa Moore
Graduate Research Assistant,
Human and Social Dimensions of Science and Technology
Arizona State University
sharlissa.moore@asu.edu

Frank Niepold
Education & Outreach Specialist,
Climate Program Office NOAA
frank.niepold@noaa.gov

Sally O'Connor
Directorate for Biological Sciences
National Science Foundation
soconnor@nsf.gov

Rae Ostman
Director of National Collaborative Projects
Sciencenter
rostman@sciencenter.org

David Rabkin
Director, Current Science and Technology
Boston Museum of Science
drabkin@mos.org

Simil Raghavan
Associate Program Officer
National Academy of Engineering
sraghavan@nae.edu

Laura Sasso
Graduate Research Assistant
Department of Science, Technology, and Society
University of Virginia
ljs9q@virginia.edu

Jen Schneider
Assistant Professor of Liberal Arts and International Studies
Colorado School of Mines
jjschnei@mines.edu

Tamara Shapiro
Ledley Senior Scientist
TERC
Tamara_ledley@terc.edu

Toni Shope
East Regional Director
The Alliance for Innovation
tshope@transformgov.org

David Sittenfeld
Forum Program Manager
Boston Museum of Science
dsittenfeld@mos.org

David Slutzky
Research Associate Professor of Urban and Environmental Planning
University of Virginia
slutzky@virginia.edu

Karl Smith
Professor of Cooperative Learning in Engineering
Purdue University
smith511@purdue.edu

Walter Staveloz
Director, International Relations,
Association of Science-Technology Centers
wstaveloz@astc.org

Laurens van der Tak, P.E., D.WRE
Vice President - Sr. Principal Technologist
Water Resources & Ecosystem Management
CH2M HILL
Laurens.VanderTak@CH2M.com

Cyndi Trang
President,
University of Marymount Science Club
t_cyndi@yahoo.com

Jeanne Troy
Program Officer
Koshland Science Museum
jtroy@nas.edu

Bill Wepfer
Chair, Woodruff School of Mechanical Engineering
Georgia Tech/ABET
bill.wepfer@me.gatech.edu

Keith Williams
Chief Technology Officer
Navy Research and Engineering Division
SAIC
KEITH.A.WILLIAMS-2@saic.com

Dick Wright
ASCE
richard.n.wright@verizon.net

Pria Young
Mirzayan Science and Technology Policy Fellow
National Academy of Engineering
pyoung@nae.edu

Jesse Zito
Public Programs Manager
Museum of Discovery and Science
Jzito@mods.net

Climate Change and America's Infrastructure: Engineering, Social and Policy Challenges – Workshop Attendees
January 28-30, 2013

Jose Aguto
Legislative Secretary, Sustainable Energy and Environment
Friends Committee on National Legislation
jose@fcnl.org

Kira Appelhans Landscape Architect
Working Earth
kirashelly@hotmail.com

Kristin Baja
Hazard Mitigation Planner
City of Baltimore Office of Sustainability
kristin.baja@baltimorecity.gov

Charles Beck
Environmental Planner
Arizona Department of Transportation
CBeck@azdot.gov

Katie Behrmann
Programs Fellow
Museum of Science, Boston
kbehrmann@mos.org

Frazier Benya
Program Officer
National Academy of Engineering
fbenya@nae.edu

Michael Bernstein
Graduate Student, School of Sustainability
Arizona State University
michael.j.bernstein@asu.edu

Thomas Birkland
Professor and Assoc. Dean for Research, Extension, Engagement & Economic Development
North Carolina State University
tabirkla@ncsu.edu

Jeanne Braha Troy
Program Officer
National Academy of Sciences
jtroy@nas.edu

Kay Brothers
former Deputy General Manager Southern
Nevada Water Authority
kayrenebrothers@gmail.com

Katja Brundiers
Graduate Student
School of Sustainability, ASU
katja.brundiers@asu.edu

Kevin Burke
City Manager
City of Flagstaff
ssmith@flagstaffaz.gov

Elizabeth (Beth) Cady
Program Officer
National Academy of Engineering
ecady@nae.edu

Daniel Cayan
Research Meteorologist
Scripps Institution of Oceanography
dcayan@ucsd.edu

Jinqiang Chen
Graduate Student
California Institute of Technology
jcchen@gps.caltech.edu

Nalini Chhetri
Climate Science Program Manager Global
Institute of Sustainability/ASU
Nalini.Chhetri@asu.edu

Ann Marie Chischilly
Executive Director
Institute for Tribal Environmental Professionals
Ann-Marie.Chischilly@nau.edu

Shauhrat Chopra
Graduate Student
University of Pittsburgh
shc82@pitt.edu

Andres Clarens
Assistant Professor University of Virginia
aclarens@virginia.edu

Chuck Cullom
Colorado River Program Manager
Central Arizona Project
ccullom@cap-az.com

Mary Ann Curran
LCA Consultant BAMAC Ltd.
macurran@cinci.rr.com

Jered Dean
Visiting Assistant Teaching Professor Colorado School of Mines
jdean@mines.edu

Jason Delborne
Assistant Professor Colorado School of Mines delborne@mines.edu

Liz Doran
Graduate Student
Duke University
liz.bloomhardt@duke.edu

Leslie Ethen
Director of the Office of Conservation and Sustainable Development
City of Tucson Leslie.Ethen@tucsonaz.gov

Mahmud Farooque
Associate Director - DC
CSPO-ASU
mahmud.farooque@asu.edu

Kathy Freas
Global Water Resources Director CH2M HILL
kfreas@ch2m.com

Elizabeth Frederick
Principal Planner
Kleinfelder
BFrederick@kleinfelder.com

Kris Gade
Biologist
Arizona Department of Transportation
kgade@azdot.gov

Gerald Galloway
Glenn L. Martin Institute Professor of Engineering
University of Maryland
gegallo@umd.edu

Nancy Gassman
Natural Resources Administrator
Broward County Natural Resources Planning and Management
ngassman@broward.org

Jay Golden
Associate Vice Provost
Duke University
jay.golden@duke.edu

Elisabeth Graffy Professor of Practice CSPO-ASU/Lightworks
egraffy@asu.edu

Nancy Grimm
Professor, School of Life Sciences
Arizona State University
nbgrimm@asu.edu

Bjoern Hagen
Graduate Student, School of Geographical Sciences & Urban Planning
Arizona State University
bjoern.hagen@asu.edu

Rebecca Hale
Graduate Student, School of Life Sciences
Arizona State University
rlhale@asu.edu

Mohammad Halimi
Graduate Student
Virginia Tech
halimi@gmail.com

Patrick Hamilton
Director, Global Change Initiatives Science Museum of Minnesota
hamilton@smm.org

Eric Havel
Education Manager
Chabot Space & Science Center
ehavel@chabotspace.org

Joe Herkert
Lincoln Associate Professor of Ethics & Technology
Arizona State University
joe.herkert@asu.edu

Lori Hidinger Managing
Director
CSPO-ASU
lori.hidinger@asu.edu

Helene Hilger
Associate Professor Emeritus
IDEAS Center UNC Charlotte
hhilger@uncc.edu

Rachelle Hollander
Director, Center for Engineering, Ethics, & Society
National Academy of Engineering
rhollander@nae.edu

Kathy Jacobs
Assistant Director for Climate Adaptation and Assessment
White House Office of Science & Technology Policy
Katharine_L._Jacobs@ostp.eop.gov

Carly Jerla
Co-Study Manager, Colorado River Basin Water Supply and Demand Study
Bureau of Reclamation
cjerla@usbr.gov

Barbara Rose Johnston Senior
Research Fellow
Center for Political Ecology
bjohnston@igc.org

Laurna Kaatz
Climate Policy Analyst
Denver Water
laurna.kaatz@denverwater.org

Eric Kennedy
Graduate Student, Consortium for Science, Policy & Outcomes
Arizona State University
ericbkennedy@asu.edu

Julia Kerran
Graduate Student, School of Geographical Sciences & Urban Planning
Arizona State University
jkerran@asu.edu

John Kester
Graduate Student
University of Arkansas
jkester@email.uark.edu

Maudood Khan
Research Scientist
Universities Space Research Association
maudood.n.khan@nasa.gov

Gregory Kiker
Professor
University of Florida
gkiker@ufl.edu

Jonathan Koehn
Environmental Manager
City of Boulder
koehnj@bouldercolorado.gov

Dhara Kothavala
Graduate Student
Arizona State University
dhara.kothavala@asu.edu

David Lapp
Manager, Professional Practice/Gestionnaire, Exercice Professionnel
Engineers Canada/Ingénieurs Canada
David.Lapp@engineerscanada.ca

Bonnie Lawless
Program Coordinator
CSPO - ASU
Bonnie.Lawless@asu.edu

Tracey LeBeau
Director, Office of Indian Energy Policy and Programs
U. S. Department of Energy
Tracey.Lebeau@Hq.Doe.Gov

Carol Lee
Graduate Student, Applied Biomedical and Health Ethics
Arizona State University
clee124@asu.edu

Robert Lempert
Director, Frederick S. Pardee Center for Longer Range Global Policy and the Future Human Condition
RAND Corporation
lempert@rand.org

Stacy Levy
Eco-Artist sere ltd
stacy@stacylevy.com

Sam Lipson
Director of Environmental Health
Cambridge Public Health Dept
slipson@earthlink.net

Pat Mariella
Director, American Indian Policy Institute
Arizona State University
Pat.Mariella@asu.edu

Clark Miller
Associate Director and Associate Professor
CSPO-ASU/SPGS-ASU
Clark.Miller@asu.edu

Brandon Morrison
Graduate Student
Duke University
bcm10@duke.edu

Armin Munevar
Global Technology Leader - Resource Systems Management
CH2MHill
Armin.Munevar@CH2M.com

Tischa Munoz-Erickson
Social Scientist
U.S. Forest Service
Tischa.Munoz-Erickson@asu.edu

Cliff Neal
Water Resources Management Advisor
City of Phoenix
cliff.neal@phoenix.gov

Byron Newberry
Professor, Mechanical Engineering
Baylor University
Byron_Newberry@Baylor.edu

Jason O'Leary
Graduate Student, Consortium for Science, Policy & Outcomes
Arizona State University
jroleary@asu.edu

David Rabkin
Director, Current Science & Technology
Museum of Science
drabkin@mos.org

Simil Raghavan
Associate Program Officer National Academy of Engineering
sraghavan@nae.edu

Neethi Rajagopalan
Post-doctoral Researcher
Duke University
neethi.rajagopalan@duke.edu

Donna Riley
Associate Professor of Engineering
Smith College
donnamriley@yahoo.com

Jody Roberts
Director, Center for Contemporary History & Policy
Chemical Heritage Foundation
jroberts@chemheritage.org

Mithun Saha
Graduate Student
Northeastern University
saha.m@husky.neu.edu

Tom Seager
Associate Professor, School of Sustainable Engineering & the Built Environment
Arizona State University
Thomas.Seager@asu.edu

Nancy Selover
AZ State Climatologist
Arizona State University
selover@asu.edu

Katy Simon
Washoe County Manager Washoe County
ksimon@washoecounty.us

Aneka Singlaub
Climate Program Coordinator North Cascades Institute
aneka_singlaub@ncascades.org

David Sittenfeld
Program Manager, Forum
Museum of Science, Boston
dsittenfeld@mos.org

Jen Snyder
Communications Specialist
CSPO-ASU
Jen.Snyder@asu.edu

Stella So
President
Ave Solutions LLC
stellaso@avesolns.com

Susan Spierre
Graduate Student, School of Sustainability
Arizona State University
susan.spierre@asu.edu

Jennie Stephens
Associate Professor of Environmental Science and Policy
Clark University
jstephens@clarku.edu

Sandy Sutton
Executive Director
Water Infrastructure Finance Authority
ymendoza@azwifa.gov

James Svara
Professor, School of Public Affairs
Arizona State University
James.Svara@asu.edu

Wendy Tao
Associate
Cambridge Systematics
wtao@camsys.com

Pilar Thomas
Deputy Director, Office of Indian Energy Policy and Programs
U. S. Department of Energy
Pilar.Thomas@Hq.Doe.Gov

Karen Thoreson
President
Alliance for Innovation
kthoreson@transformgov.org

Thomas Wilbanks
Corporate Research Fellow,
Climate Change Science Institute
Oak Ridge National Laboratory
hallta@Ornl.gov

Wendy Woodward
Science Teacher
Show Low High School
WendyW@show-low.k12.az.us

Richard Wright
Member
National Academy of Engineering
richard.n.wright@verizon.net

Appendix C

Summary of Evaluations

Appendix C
Summary of Evaluations:
Did Workshop Presentations Enhance Participants' Understanding?

Workshop attendees and project team members were invited to complete an evaluation after each workshop to assess what information they considered useful, what other information might be useful, and whether the presentations increased their understanding of the topics discussed. The evaluations were part of a formative assessment of the project to determine whether it gathered the appropriate information to achieve its long-term goals.

Questionnaires were distributed via paper at all three workshops. Evaluation was tailored to the specific sessions, although some questions (e.g., whether understanding was increased) were similar across all three workshops. Because separate evaluations were given at the end of each workshop day the number of respondents differs across the various workshop sessions.

The responses were analyzed by the project evaluator and are summarized here by topic as presented in the report chapters.

Interactions – Defining the Problems

Of the 26 evaluations returned after the first workshop, a majority of respondents agreed or strongly agreed that the panel presentations on Interactions afforded them a better understanding of how scientific, engineering, or social interventions influence the interactions of climate, engineered systems, and society (Table A.1). In addition, most of the attendees rated the speakers (McCarthy, Golden, and Bostrom) and respondents (Herkert and Delborne) as effective, although two thought the presentations were too general.

Of the 23 capstone workshop attendees who returned the evaluations, most agreed both that they gained a better understanding of climate change, adaptation, and infrastructure vulnerability (Table A.1) and that they would use the information learned. Nine attendees listed other topics that could have been usefully included:

- More information on ecosystem services; more illumination of human connection to natural world and identification of areas/connections that, if damaged, are the biggest threat.
- Climate change affects the economy/society not just through disasters but also altering the ecology, thus affecting ecological services. We need a discussion on increasing climate change discourse in this area.
- How do international political tensions affect climate change?
- Perhaps a presentation by a government/federal policymaker to provide insight into issues related to how the government sets policy and collects info to do so.
- Rural concerns – thought Patricia Mariella did address this to some extent.
- Engineering solutions that have been implemented and have been successful for adaptation.
- More thorough explanation of exactly what the smart grid is.
- What do the educational and outreach materials for passing this information along to the public look like?
- How to connect/start up a stakeholder network on climate adaptation knowledge access?

TABLE 1 Percentage of respondents indicating they "Agree" or "Strongly Agree" that they gained a better understanding of the problem.

This session gave me a better understanding of:	%
How **scientific** interventions influence the interactions of climate, engineered systems, and society.	**65**
How these influences might affect the success of programs and recommendations.	**69**
How **engineering** interventions influence the interactions of climate, engineered systems, and society.	**65**
How these influences might affect the success of programs and recommendations.	**77**
How **social** interventions influence the interactions of climate, engineered systems, and society.	**62**
How these influences might affect the success of programs and recommendations.	**62**
Climate change, adaptation, and infrastructure vulnerability.	**87**
Ways in which climate change, adaptation, and infrastructure vulnerability are interconnected.	**78**

Interventions Examining the Range of Sociotechnical Responses

After the first workshop most of the 23 respondents indicated that they gained a better understanding of the social dimensions and potential consequences of adaptation, mitigation, and geoengineering in response to climate change (Table A.2). A majority also rated the speakers (Rubin, Kepke, Daniel, and Robock) and respondents (Johnson and Slutzky) as effective. Although one attendee commented that the talks were too general, others commented positively on the session:

- Really appreciated this session—Ed [Rubin] & Jackie [Kepke] are "in the trenches," so to speak, very conversant on climate/engineering/policy issues.... This session really crystallized things for me.... Alan Robock's presentation was also great. Maybe geoengineering shouldn't be in the grant...but the technical solution is so sexy for engineers? It does seem shortsighted of us not to address it in some way, especially if it happens without us.
- Jackie was a very effective, powerful speaker. Hopefully she will become an exemplar for how CC-relevant engineers ought to be.
- Alan's geoengineering myths need to become more visible and relevant, especially to engineers that invoke geoengineering as a technical fix to climate change.

After the talks by Munevar and Lapp at the capstone workshop, a majority of attendees agreed that they gained a better understanding of engineering perspectives on climate change and infrastructure problems as well as the ways in which engineering, climate change, and infrastructure resilience are interconnected (Table A.2), although some thought the discussions should have presented examples of successful engineering solutions.

Based on presentations about water resources, including two (Kiker and Lempert) on sea level rise and storm surge and a panel on the Colorado River (Jerla, Brothers, Mahmoud, and Neal), a majority of the 21 respondents reported a better understanding of sea level rise, storm surge, and resources available in and around the Colorado River (Table A.2) and indicated that they would use the knowledge acquired in the sessions in the future. One respondent specifically mentioned "reverse planning" and another the "cluster analysis and relationship/contrast between vulnerability analysis and systems analysis." Attendees also thought it would have been useful to include human justice and land use issues related to the Colorado River, presentations from different parts of the United States, and a "wider range of infrastructure (e.g., the focus was energy and water/flooding)."

TABLE 2 Percentage of respondents indicating they "Agree" or "Strongly Agree" that they gained a better understanding of the problem.

This session gave me a better understanding of:	%
Social dimensions of adaptation in response to climate change	**65**
Potential consequences of adaptation	**74**
Social dimensions of mitigation in response to climate change	**70**
Potential consequences of mitigation	**61**
Social dimensions of geo-engineering in response to climate change	**78**
Potential consequences of geo-engineering	**91**
Sea level rise and storm surge	**81**
The resources in the Colorado River	**61**

Cross-Cutting Themes

Based on the Cross-Cutting Themes panel at the first workshop, a majority of attendees agreed that they gained a better understanding of justice, sustainability, governance, trust, and public engagement in relation to climate change education (Table 3). The speakers (DesJardins, Thompson, and Moser) were viewed as effective, although one person commented that the presentations needed to link more directly to "climate change or engineered systems or both" in order to apply more directly to the project.

The capstone workshop included two panels relating to cross-cutting themes, and a majority of respondents again indicated that they gained a better understanding of these issues in relation to climate change, adaptation, and infrastructure vulnerability (Table 3) and would use the information they had learned. Despite the positive ratings, some respondents listed information they felt was missing from the discussions:

- A session exploring social justice conceptualizations (in the context of ways of life). More on how clients and communities work with engineers to address problems.
- A little more of a world-view might be nice if there are any, like Australia's aborigines having a share of water rights.
- More on environmental justice/equity issues.
- The governance panel/discussion could've used more examples of cases of the social and political dynamics of how these vulnerabilities were developed in the first place; and how we deal with the structural thinking patterns that may limit adaptation and rebuilding sociotechnical systems.

TABLE 3 Percentage of respondents indicating they "Agree" or "Strongly Agree" that they gained a better understanding of the problem.

This session gave me a better understanding of:	%
Justice in relation to climate change education	**63**
Sustainability in relation to climate change education	**54**
Governance in relation to climate change education	**54**
Trust in relation to climate change education	**67**
Public engagement in relation to climate change education	**67**
The policy and governance challenges and strategies for assessing the problem of climate change, adaptation, and vulnerability	**65**
The engineering, justice, and human rights challenges inherent in climate change, adaptation, and vulnerability	**70**
How engineering, climate change, adaptation, policy and governance, justice, and human rights interconnect	**87**

Potentials in Formal Engineering and Science Education

Of the 23 attendees at the second workshop who returned the evaluations, the majority indicated that the session on effective interventions in undergraduate engineering education had enhanced their understanding of both diffusion of educational innovations and associated challenges, opportunities, and barriers, although they came away with a clearer idea of institutional barriers than local/state or national/federal barriers to integrating climate change and engineered systems in engineering curricula. Most also agreed that they had a better understanding of K–12 science standards and learning progressions, but were less clear about how engineering or climate change fits into K–12 curricula (Table A.4). A few respondents said they had trouble relating information presented in the K–12 engineering session to the project.

Following the capstone workshop, a majority of the 15 respondents indicated they had a better understanding of the challenges of educating engineers about climate change, adaptation, and infrastructure vulnerability (Table A.4) and that they would use the information learned in the session, although some were unsure how to translate the information into practice. Two respondents found Riley's presentation and initiatives particularly helpful.

TABLE 4 Percentage of respondents indicating they "Agree" or "Strongly Agree" that they gained a better understanding of the problem.

This session gave me a better understanding of:	**%**
Diffusion of educational innovation	**96**
The potential challenges of integrating climate change and engineered systems into the engineering curricula	**96**
The potential opportunities of integrating climate change and engineered systems into the engineering curricula	**83**
The potential institutional barriers to integrating climate change and engineered systems into the engineering curricula	**87**
The potential local and/or state barriers to integrating climate change and engineered systems into the engineering curricula	**35**
The potential national/federal barriers to integrating climate change and engineered systems into the engineering curricula	**35**
How engineering curricula are used in K–12 education	**48**
The potential challenges for the topics of climate change and engineered systems in the changing science standards for K–12 education	**57**
The potential opportunities for the topics of climate change and engineered systems in the changing science standards for K–12 education	**43**
The idea of a naturalized philosophy of sciences	**17**
How science and engineering practices can be integrated into science standards	**74**
Learning performances and progressions in relation to science standards	**70**
The challenges of educating engineers about climate change, adaptation, and infrastructure vulnerability	**87**

Potentials in Informal Engineering and Science Education

Most respondents at the first workshop said they gained a better understanding of the role of science and technology centers in education efforts related to climate change (Table A.5) and found the session useful and engaging. For a majority of the 15 respondents, presentations by representatives of the Boston Museum of Science, the Marian Koshland Science Museum, the Science Museum of Minnesota, and Chabot Space and Science Center yielded a better understanding of informal science and engineering education as well as ways science museums and formal education institutions can work together. Respondents also indicated that they would use this knowledge, although some commented that they were unsure how to translate it into action (Table A.5). There were several suggestions of other information that could have been included:

- Other types of informal education including nonprofit environmental education organizations, environmental learning centers, government agencies like National Park Service, USDA Forest Service Interpretation.
- Historic preservation and architecture perspectives were missing from the conversation and at the whole conference.
- Inspired by Science Center discussion, but would like to consider who these audiences are, who can make it to a museum? Not the vulnerable populations. Cultural inclusion in today's world is mandatory.

The panel presentation from artists (Roberts, Appelhans, Levy, and Mosher) led a majority of respondents to agree that they gained a better understanding of engaging the public in climate vulnerability and adaptation (Table A.5) and would use the information they learned. As with other sessions, some respondents commented that including ways to apply the information to their work would have been helpful, and one stated specifically that "best practices for stakeholder gathering/engagement [such as] how to recruit, structure and encourage collaboration… could have been included in "Engaging the Public" panel." Another stated that "it would be better if there is a topic in the public's sense of value towards the climate change from multi-perspectives including people from different class levels," but one respondent "appreciated the session on arts/design – thought it was a helpful way to bridge the techno-specific focus of engineering and the human-dimensions focus of human rights/justice, etc."

TABLE 5 Percentage of respondents indicating they "Agree" or "Strongly Agree" that they gained a better understanding of the problem.

This session gave me a better understanding of:	**%**
The role that science and technology (S&T) centers play in the educational community	**91**
The institutional strengths of S&T centers in communicating multifaceted information	**61**
The role that S&T centers can play in local and regional outreach	**87**
How S&T centers will engage the general public and school-aged audiences in the topic of climate change, engineered systems, and society	**96**
Informal science education opportunities in climate change, adaptation, and infrastructure vulnerability	**100**
How informal science education and engineering education can work together to increase knowledge of climate change, adaptation, and infrastructure vulnerability	**67**
Engaging the public in climate vulnerability and adaptation	**76**

Perspectives of Engineering Professional Societies, Business and Industry, Local Government, and Native Americans

The second workshop included perspectives from both engineering professional societies and industry representatives. Of the 16 attendees who returned the evaluations, a majority agreed that they had a better understanding of the educational priorities of the societies represented (Table A.6), although comments indicated that Lapp (Canadian Standards Association) was the most helpful and informative. A majority of those respondents also indicated that they had a better understanding of what employers expect from graduates in terms of underlying engineering principles and engineering skills, although they did not gain as much knowledge related to expectations of engineering experiences. A majority also gained understanding of whether employers recognize and how they are working to address issues concerning climate, engineered systems, and society (Table A.6). Comments indicated the session was helpful and suggested topics for further discussion:

- Tension between student technical and communication skills – desirable for employers vs. social/ethical training that can't apply if it isn't profitable?
- Educating corporations to view justice beyond corporate social responsibility or philanthropy.

- The challenge is always that employers desires are often contradictory or difficult to translate to educational experience (e.g., they must be technically excellent but must also be great communicators but don't give them too many weird projects).

A majority of the 21 respondents to the capstone workshop panels also agreed that they gained a better understanding of local government solutions to and Native American perspectives on climate change and infrastructure vulnerability (Table 6) and would use the knowledge moving forward. Comments indicated that 7 respondents found the Local Government Solutions and another 2 found the Native American Perspectives panels particularly helpful. Comments also suggested topics such as "Native American traditions of resource management that seem to be part of their culture as a model for thinking about sustainability in the engineering world" for future consideration.

TABLE 6 Percentage of respondents indicating they "Agree" or "Strongly Agree" that they gained a better understanding of the problem.

This session gave me a better understanding of:	%
The educational priorities of ABET in terms of climate, technology, and society	**69**
The educational priorities of ASCE in terms of climate, technology, and society	**75**
The educational priorities of the Canadian Standards Association in terms of climate, technology, and society	**94**
What employers expect from engineering graduates in terms of underlying engineering principles	**75**
What employers expect from engineering graduates in terms of engineering skills	**63**
What employers expect from engineering graduates in terms of engineering experiences	**44**
Whether employers recognize the issues concerning climate, engineered systems, and society	**88**
How employers are working to address the issues concerning climate, engineered systems, and society	**69**
Local government solutions to climate change and infrastructure vulnerability	**100**
Native American perspectives on climate change and infrastructure vulnerability	**76**